Fundamental Intelligence, Volume I: AI as a Label

Rethinking Natural and Artificial Intelligence

Ali Magine

Fundamental Intelligence, Volume I: AI as a Label
Rethinking Natural and Artificial Intelligence

Published in the United States
ISBN: 978-1-7365633-1-1 (paperback)

First Edition 2022

Fundamental Intelligence, Volume I:

AI as a Label

To My Son,

Benjamin

"All of humanity's problems stem from man's inability to sit quietly in a room alone"

Blaise Pascal
(1623-1662)
French Polymath

Contents

Chapter 0

Introduction

The Case for This Book

By now, almost anyone on the planet with access to the internet has at least heard of Artificial Intelligence (AI). The large sum of capital that has been spent under its banner, and the interesting results in both industry and academia, has naturally attracted an enormous crowd to want to learn more or even specialize in this field, and many to be expert at it already, or at least so they say. There is no stopping it from being thrown around as a fancy label. That is AI as a label. Although that makes up for the title of this volume, it is not at all the purpose of the book to hammer on the use or misuse of the phrase. That ship of resisting the term has long sailed.

Truth be told, I, among many, would like to be able to call myself an *intelligence scientist*. I cannot, nor can anybody really, there is no such science. Not yet. We can ask though what it would take to establish it. I believe this book can serve as the very first step towards that by at least clarifying the problem statement better. I feel very fortunate to finally get to this point and find the time to write about it with the breadth and depth it deserves. It is based on years of observations from both outside and inside the field of AI, as well as years of reflecting on my own thinking and compiling the results.

At the rate it's going, soon there will be more publications on AI than any other discipline in history. There are already a lot of books out there on AI, from textbooks and practical guides to science fiction. So, what is exactly missing? Well, a cross-disciplinary big picture of the foundations and fundamentals of AI is direly missing. Filling this gap at the very least could be of conceptual help to many.

The attempt here is not just about painting a big picture, but also to argue and explain how the existing foundations are not satisfying if we are serious about intelligence. The current dominant mindset and philosophy of AI are built on shaky grounds (for historical reasons). In the first volume (that is this book), we only get to motivate this statement, by casting doubt on existing philosophies. Questioning historical assumptions alone, sets the content of the book radically apart from any other text on AI, let alone the introduction of net-new ideas and perspectives.

Artificial intelligence is obviously too important not to be understood deeply and paid attention to. Endless debates, confusions, and misunderstandings in and around AI can be a promising sign of inevitable progress to come. It can also be a strong sign for the absence of proper foundations, and the book argues that it is. Perhaps this is what should be scary about AI, not the meaningless portrayals of super-intelligence. Without having a fundamental big picture of AI, we cannot fully stop asking wrong or meaningless questions.

An important message implied in this Volume is that it should be front and center the admission that we don't understand intelligence. It shouldn't be sidelined simply because this admission is a necessary step towards paradigm-shifting breakthroughs. We're going to ask the toughest questions and unlike any other text, we are not going to pretend that anyone knows or is even close to knowing the answers to them, all the while taking into account recent advances.

Contributions of The Book

Abstract

The contribution of the book is to bring a collection of arguments to make the case why no one can currently give an answer to the problem of intelligence without revising the foundational philosophies of AI. It makes the case that any attempt to produce an answer will result in a non-answer (or a hack with limited utility) unless it revises foundational philosophies and restates the problem of intelligence. It makes the case for why we need a problem statement for intelligence that is unlike the word intelligence is not prone to a diverse taxonomy. This work introduces a new hypothesis called fundamental intelligence, justifies it, and builds upon it to reformulate the problem of intelligence. This can provide insights that can assist the community in building benchmarks for benchmarks and conduct more intelligent research on intelligence!

Better Problem Statement

The goal of most AI researchers is to "solve intelligence", as Google Deepmind puts it. If you ask about the problem statement though, the best answer you can get, from whoever labels their work "AI", is along the lines of, we want to build more autonomous/useful systems and the problem is that we don't know how to build these machines currently. Quite naturally then, people focus on building things for very specific tasks. As a result, we get an ecosystem of systems that only attempt to beat performance benchmarks on narrowly defined tasks or domains.

This ecosystem cannot be expected to evolve to have a benchmark for benchmarks that could define and measure intelligence or whatever the ultimate concept should be. That should be obvious simply because it was never set out to be the goal in the first place. Achieving effectively broad intelligence (to avoid using the term "general intelligence") has only been the HOPE while the PLAN was to build "narrow intelligence", super competent in specific tasks. Why would anyone expect an ecosystem of research programs to build more competent machines that lead to more than a collection of independent and task-based benchmarks? The false hope here is that this problem/issue can disappear on its own, meaning we can continue with trial and error research programs and somehow the general AI emerges if we work on it long enough, build bigger systems, combine them in some hybrid, or define some ad hoc meta-benchmark, etc.

If we are serious about "solving intelligence", shouldn't our first step be to define or state the problem better? This quote from Einstein regarding what he believes to be the optimal problem-solving path is quite fitting here:

> *"If I had an hour to solve a problem and my life depended on the solution, I would spend the first 55 minutes determining the proper question to ask, for once I know the proper question, I could solve the problem in less than five minutes"*

According to Einstein our current approach to solving intelligence is suboptimal (at best). Our reaction to this reality seems to me to come up with excuses, the biggest one of which is to blame the concept of intelligence itself! We say that intelligence is such an overloaded term. There is ample evidence that there are many forms of intelligence and that many have tried to go after defining intelligence but got wrapped up in never-ending philosophical debates. Promotion of this excuse goes straight back to Alan Turing, the founding father of AI.

The truth is ignoring the problem of stating the problem of intelligence and rushing to build something, that can imitate intelligence, is easier. It's much harder to figure out what the right questions to ask are and what the right problem statement is. The hardness of finding the right problem cannot be underestimated. People have indeed tried to define intelligence time and time again but it proves too hard to arrive at something useful as well as unifying. Pursuing such a definition quickly fatigues us and we move on to build something useful for the business tomorrow. However, this reason is also an excuse. Maybe we should get past defining the word intelligence, but we shouldn't get past finding the right problems to substitute it with. This book's target is to replace it with an abstract parent concept, something more fundamental, that subsumes the variety of behavior and phenomena that we would like to attribute intelligence to.

Nature is made of information to our best current understanding. The book argues that information processing is an equally intrinsic part of nature. Thus, the term "fundamental" is also justified in that sense, in resetting the problem of intelligence with roots in natural sciences, or physics, for that matter.

Post-Turing (Research) Programs on AI

Alan Turing appears to have anticipated almost everything we got now, at least at an abstract level, and quite ingeniously set us up to go after them. The style of thinking he introduced has shaped computer science till now. The most pervasive of all is to replace a problem of interest with another problem which we may solve easier. This technique sits at the core of theoretical computer science since its dawn till now. He did just that with AI too, namely, replacing the "problem of intelligence" with problem of imitating or fooling human intelligence.[1] Adopting Turing's philosophies has taken us astonishingly far, in computer science and AI. However, a mysteriously hard problem (not irrelevant to intelligence) is how to know when an idea or philosophy is no longer good or relevant. The book argues that it may be time to question whether the implicit foundational philosophies (in short, we call them "Turing philosophies") are outliving their high-utility period. In particular, we suggest that falling short on investigating the following areas should be traced back to Turing's founding agenda for the field and the fact that we collectively seem to have been mesmerized by it till now:

1) Imitation is sufficient— Replacing the problem with another problem is a fine technique and can even serve to define the original problem statement better. Though, that's not

---

[1] Turing, A. M. "Computing Machinery and Intelligence." *Mind* 59.236 (1950): 433-460.

always easy to do and even less easy to tell whether we have succeeded in doing so. This is the case with Turing-based or imitative AI research. It is an anthropomorphic agenda for intelligence to begin with.[2] Chapter 3 introduces the notion of a fundamental bias of our brain which argues that when it comes to intelligence, we need to be even more cautious with anthropomorphizing than we usually need to be.

That's a bias so powerful that has made AI somewhat of an emotional field driven by our intensely strong and yet natural desire to see or at least wonder about truly intelligent machines that could do anything we do and more. That's been almost like a virus that everybody's got, and it's been blindfolding us since Turing.

2) Scale solves it— Turing believed that with bigger scale some magic can happen, in analogy with a nuclear bomb that the cross-over to super-critical mass suddenly gets the job done.[3] Imagine if the Manhattan project didn't have sound theory and calculations of super-criticality thresholds. We don't think Turing suggested to "solve intelligence" without having a theoretical understating of how the magic is supposed to happen. Nevertheless, it seems to have induced the hard-to-tame fallacy that scale alone can, or maybe able to, solve it.

3) Goals and objectives can be arbitrary— There is an absolute absence of any talk of what the goals or objectives of a machine should be in foundational texts. This has silently led to an overwhelmingly widespread (trivial sounding) belief that goals and objectives are not within the scope of intelligence or intelligent systems, but rather something that needs to be supplied to them by the external world or the engineer of the system. This has, in turn, licensed adoption of the philosophy that something like a "general intelligence" exists, in the sense that you can first come up a general intelligence and then have it achieve pretty much any goal you give it or it chooses. In a post-Turing philosophy, goals should not be treated as a trivial concept, from where they come from to how they are intertwined with intelligence. Brining significant attention to this topic is quite hard in the world of AI,

---

[2] Among the greatest philosophers, Daniel Dennett calls it Disneyfication of AI, referring to the imitation-based foundations for AI. See Daniel C. Dennett, "What Can We Do?" in Possible Minds: 25 Ways of Looking at AI, J. Brockman, ed., 2019.

[3] Ibid, 1.

which itself is the evidence of us being still deep in the mesmerizing world that Turing ignited for us.

4) Data and decisions aren't the focus— What is meant by data here is whatever that comes as input to an AI, and decisions are whatever's done with the output of the AI, downstream from it. Data and decisions were left out of the core of foundational topics in AI, similar to the status given to the topic of goals, i.e. not related to intelligence. This has led to a collective lack of emphasis on them to this day. This is a complete bias, that is hard to even recognize, i.e. our bias in not realizing the importance of the topic data and decisions and assuming independence for them from what intelligence is.

These philosophies do have many problematic, often indirect, consequences for the real-world practice of AI. However, our emphasis here is NOT to go after discussing or addressing them one at a time. That would be just going down the road of "curing the symptoms". Let me quote Einstein once more: "We can't solve problems by using the same kind of thinking we used when we created them."

Curing-the-symptom research programs tend to result only in hacky patch-work or temporary solutions, as opposed to a principled and fundamental fix. We are advocating, instead, for the need to chase problems down to their deepest roots, which we attribute to our multi-generational unquestioning and unconscious adoption of, as we said, Turing philosophies. We have been living and breathing these philosophies for generations. Therefore, we should all assume that we are all intensely biased here, simply because that's the only world we've been witnessing, and more importantly, still are.

We argue that a necessary condition for coming up with a bona fide problem of intelligence is to revisit those foundational philosophies. An explicit realization can provide deep links between seemingly diverse (research) programs. Therefore, an explicit and formal recognition, not only helps us pay more attention to foundational issues but is also beneficial to existing programs. What this book is yelling is that course-correcting cannot be done without a serious engagement in philosophy of intelligence as that's where we have gone wrong.

A post-Turing research program is one that builds on philosophies that explicitly deviate from the above set. Be it so, in the words of Alan Turing, you may describe the contents of the book as only "recitations tending to produce beliefs"! I will certainly keep pulling on this thread, to make stronger cases and develop the ideas further, but certainly this is an effort that needs contributions from many other minds to form hard bones as a program. Therefore, part of the mission for this book is to provide the minimum sufficient material for any interested party to be able to join the discussion and contemplate a "post-Turing" research on intelligence.

Content for Cross-Disciplinary Engagement and Understanding

The book focuses on issues at the intersection of philosophy, science, technology, and humanity. The goal is to provide original and refreshing perspectives while making the text as self-contained as possible. It should help us form big-picture intuitions, see specific points of view as stemming from just one self-consistent holistic philosophy, rather than multiple diverging theories.

In parallel, I am attempting to take an initiative to combat some cross-discipline walls without resorting to much technical language in order to reach a wider audience. I strongly believe the current atmosphere we're living in necessitates such a text to shape a discourse that involves many disciplines and only later attending to full academic rigor and polish. I believe a conversational tone can serve best here and that's what I'll adopt throughout the book.

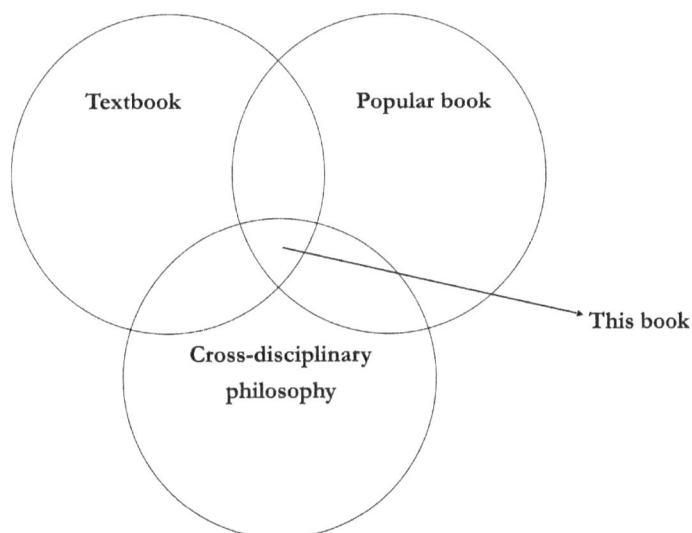

The other factor to keep in mind here is that the offered content may be highly controversial for several reasons:

- AI and associated topics are highly controversial in general,
- Revisiting foundations present unconventional views, by definition.
- Going cross-disciplines always risks making some folks feel you didn't do them justice while stepping on their turf.

Necessity of an Interdisciplinary Approach

AI is an intrinsically multifaceted topic, and as such, demands going cross-disciplines. For instance, to humanities, philosophy, psychology, and physics. Among which the link to physics may be the least obvious, a topic that we must and will explore later in the book.

Deep issues are almost always multifaceted and interdisciplinary. Any inherently multidimensional treatment is better than any single-dimensional one. In that spirit, it's best for us to be problem-focused instead of discipline-focused or going based off of who is specialized in what. That also necessitates covering the most controversial issues rather than shying away from them. I cannot resist borrowing a thought from Richard Feynman[4] here pointing out that: nature doesn't read textbooks on various disciplines one at a time, on physics, chemistry, geology, cognitive science, social sciences, and alike. Nature is nature!

There are many people who are interested in reaching a science of intelligence and are currently dispersed in different communities. The hardest part may be to bring everyone together as we need both non-trivial depth and non-trivial breadth to focus on the problem holistically and solve it, not unlike what a product group does working under the clock to release a product, with vastly different sets of expertise sitting around the table.

Good new philosophies can uniquely help with that. Building a new science like building anything starts with building a foundation, and the foundation for any science is some philosophy, even if unspoken.

Who Is This Book for and What May You Get out of It?

I am writing for a broad audience, for all those who are passionate about the prospects of replicating some aspects of their own mind in useful machines and all the exciting things that would come with it. But equally so for those who are fed up with what's out there and are searching for a deeper text on intelligence. Whatever your current profession, if you have been looking to go at AI from a first-principle perspective, I have written this book for you! There really hasn't been any place for it, not even in the philosophy of AI. Nothing of relevant substance exists on it. That's what I hope to change a bit with this book.

---

[4] American physicist (Nobel laureate) and arguably the greatest explainer of complex science that the world has ever seen.

Our philosophy here is that it shouldn't be complicated, period. If it is, it only signals a lack of understanding. We'll work hard to bring the lack of understanding to the light of the day. Your background or education shouldn't matter. Only your interest in the topic matters. It's written so that anyone could follow. It gives you the insights that you'd otherwise have to get from knowing a lot of math, reading thousands of academic papers, and putting them together and thinking about them deeply.

I will refrain from using any quantitative formulas or descriptions. Instead, I'll stick with qualitative descriptions and what-it-all-means type of language, unlike all technical books on AI which are focused on detailed technicalities. That doesn't mean that the content of the book is not technically sound, quite the contrary, it's written to attract highly technical minds in AI as well, rather than turning them off.

Keeping it interesting for both the very beginner and the very advanced made the book quite challenging to write. As a result, the presented content is both broad and deep, and therefore, somewhat dense. It would best serve those AI enthusiasts or professionals who want to

- See how all the topics and methods relate to each other and to build a concise conceptual map, without the need to dive deep into the weeds of lots of technical papers, which is understandably impractical for many.
- Get more foundational depth around AI, and separate out what really matters.
- Just build some good judgment, intuition, and understanding without the big words and the unnecessary technical language of the esoteric treatments.
- Apply that depth to any of the ongoing debates you may find yourself a part of, articles you read, or the research that you conduct.
- Draw a more careful and educated guess as to where we are and how far off from certain capabilities in AI.
- Generally, gain a different perspective in many relevant AI topics and a sobering perspective on how much we really don't understand.

You don't learn how to write AI programs in this book. There are plenty of other resources for that. Instead, you may learn what to make of things or how to think of them differently. If you can pair some introductory texts on AI or machine learning with this book, you can get a lot more out of it and can likely appreciate the big picture simplifications offered here even more.

We are all in some form of Ivory tower, and often, getting out of it can be a helpful thing to do. I am optimistic that the book could help with that. If you walk away looking at intelligence differently, I will think that I have done my job right. If it also enables or inspires you to pick up a different direction, design, model, or implement things a bit differently than you'd do otherwise, that'd be an excellent plus.

Most of all, I hope to inspire early students, beginners, or newcomers to the field of AI to see the wealth of unknowns and shaky philosophies; to see possibilities other than committing to pre-existing tribes and schools of thought; to see that taking a brave break from short-term-oriented spotlight-chasing thinking may be necessary to arrive at breakthroughs.

How the Content Is Organized

This book is only the first volume of three. This first volume includes all introductory materials and is more about explaining where we are in the grand scheme of things, separating what we know from what we don't. In this volume, we do not get serious about defining the words intelligence and artificial, we just question them. We treat AI as just a label and focus on explaining what it is that people do under that label. In other words, we stay inside the box, look at what's in it and question it a bit. That will give sufficient materials for the reader to be able to reason about why no one has "the answer" to the challenge of creating "true AI". In the next volume, we get even further, we'll see any answer that we could give now will also be wrong unless we revisit foundational philosophies and reformulate a different problem.

At the end of this volume, we present a fictional tale written just to reset our minds a bit and clear the slate to be more open-minded ahead of the second volume, in which we get totally out of the box. We do get serious about the problem of defining intelligence. That means a total rethinking of natural and artificial intelligence. It also justifies the subtitle of this volume to be the first step towards rethinking intelligence. Finally, in the third volume, we make the case for a proper science of intelligence.

Chapter 1

Background — Where Are We?

The World of AI

Artificial Intelligence, AI! The topic that you can't escape from. Every other day it shows up in a big headline somewhere. It's always hard to quantify how much of a hype some trend is while you're living through it. As for the AI trend, the word started reappearing on the public sphere roughly around 2013, reached peak hype in 2016-2017, and has been almost steady since then in terms of coverage and the usage of the phrase.

Not surprisingly, 2017 was also the year with the largest total number of investment deals in "AI" startups. 2618 deals for a total of $15B globally.[5] There have been massive governmental investments in AI too. The Federal Government funded $1.5B of unclassified R&D in AI in 2017, in addition to substantial classified investments.[6] In 2018, DARPA alone announced a minimum of $2B investment in its "AI next" campaign to fund programs paving the way for next-generation AI technology. Yet, these budgets are an order of magnitude smaller than internal corporate investments in AI R&D, which too has soared in recent years. In 2017, big tech companies such as

---

[5] stateof.ai
[6] 2018 public report nitrd.gov

Alphabet and Amazon who have expressed core emphasis on AI research, spent $13.9B and $16.1B, respectively, in R&D efforts.[7]

Meanwhile, the pace of growth of academic research in AI and closely related fields has been unprecedented in the relatively short history of mankind. Just 6 of the major technical AI conferences alone now hold almost 30 thousand participants every year. This is as large as it has ever gotten in any scientific society.[8] Each of these conferences witnessed a 20% rise in participation year over year since 2012.[9] The much-celebrated conference NeurIPS now gets sold out in minutes after opening for registration!

Of course, these numbers may seem dwarfed next to AI's potential contribution to the global economy over the next decade, which is in tens of trillions of dollars. Contribution by AI to global GDP in 2030 is estimated to be $15.7 trillion. [10]

Although it's hard to overestimate the impact of AI on businesses, there are many other aspects to AI that make it much more than yet another popular technology. Those are impacts that AI has or will have on humanity and every one of us. From what it means to be a human in the age of AI to how we do anything. So the nature of its importance is highly multi-faceted: philosophical, governmental, safety-related, social, scientific and so on, not just economical. On the other hand, it is the platform that, at some point, will finally force us to consider all these seemingly separate facets of modern life in some delicate inter-connected theme.

Even from a purely technological point of view, developing AI to be deployed out in the wild breaks down almost all of our traditional paradigms. Less than a decade ago, the mindset that "software is going to eat the world" was popularized, and in a sense it's still happening, the idea that every company has to have a software side. It's true that AI is basically just software too but it's a kind of software that has the potential to change the meaning of modern software. Director of AI at Tesla, Andrej Karpathy, calls it software 2.0 to encourage everyone to think deeper about the influence of AI on software and software development in general. Some say "now AI is going to eat software that is eating the world"! That is because increasingly the non-AI portions of a software stack are going to be developed using data and AI methods.

---

[7] 2018 report by AI index

[8] Only mature large societies such as chemistry, biology or neuroscience see such levels of participation.

[9] aiimpacts.org

[10] PwC research

You can hardly find any topic that people have not already written about the potential of AI for or have not discussed the impact of AI on. Many fields have seen AI methods finding their way into their respective agenda. They have already been finger-tip-touched by AI. So this all fuels the hype and the excitement in a legitimate way. However, wild speculations are not in any shortage as is the case with any buzzword. What are some of these speculations and why do I call it a buzzword? That will become clear shortly. But given its multifaceted importance, AI is not an ordinary buzzword, rather a complex buzzword worth paying attention to. Here's the main question that everyone in a hype cycle would appreciate the answer to: how do you separate the real part from the imaginary?

Let's look at current approaches by three stereotypes: a technical practitioner, a non-technical professional, and a scientist. First, the technical practitioner, who looks under the hood of any current AI system in which s/he basically finds nothing but some form of Machine Learning (ML)[11] and may conclude that AI is pretty much nothing but a buzzword at the moment. The widespread joke used to be that ML is most often written in Python, while AI is only written in PowerPoint. Second, the non-technical professional, who looks at current capabilities, rate of progress, and similar variables, gets excited and makes semi-linear projections to make sense of the hype. Third, the scientist who is mostly interested to understand what's possible rigorously, looks at some challenges against ideal scenarios and makes projections that way. The approach in this book is going to be closest to that of the scientist! At least, that's what you'd expect from the first step in a fundamental rethinking of it all.

Outside of the sphere of people whose jobs cross-path with AI or those who are knowledgeable about AI, there seems to be a rather surprising perception among many that AI, to a great degree, is solved already. Impressions like "Look, Google has this AI that calls around and schedules appointments for you and so on... they pretty much have AI nailed already, right?".

On the other hand, no one truly knowledgeable in AI would tell you it's anywhere close to being "solved". How can there be such a big disconnect with the public's perception? There are only two sources of information available to a member of the general public when it comes to AI. One is their news media publishing catchy headlines, and the other is their direct experience as a consumer.

---

[11] Roughly speaking any piece of software that uses data to increase its performance, is considered an ML system.

Headlines are occupied by attention-grabbing systems like AlphaGO[12], or technologies with immediate massive consequences such as self-driving cars, and humanoid robots that are psychologically conflicting for people on multiple levels; job-wise, safety-wise, morally, and so on. Pair that with consumer experience of impressive voice recognition, natural language, and dialogue systems, along with recommender systems that seem to know a bit too much about you. Sparkle on top Hollywood movies on AI, plus tweets and statements from trusted public figures, warning us to be all scared of AI. Does it now seem surprising that people think that AI is almost solved?

According to a global survey of online consumers,[13] understanding of AI and its current status vary a lot among people except when it comes to concerns for the loss of jobs and cybercrimes. Both under-appreciated topics by AI technologists.

Now, suppose a curious individual decides to pull back the curtain a bit on what all this AI hype is about. There are two approaches to take. One is absolute, the other is relative. In the absolute way, you would try to think fundamentally by looking at what the theory of AI would say and predict. Well, that wouldn't work because there is no such theory of AI. In the relative way, you just track the source of excitement and hype, and what has verifiably changed RELATIVE to the past, say 10 years ago or so. This way, you are bound to encounter a dominant narrative, a story that's been pushed to be front and center, repeated and echoed many times over. And that's the story of the so-called recent "AI revolution", which I referred to as the "AI trend" earlier. You ask why revolution? what's behind it?

Almost certainly the story you'll be told starts with the landmark 2012 paper[14] which reports a computer program making almost half the mistakes of the runner-up program (previous state of the art) in recognizing the category of objects in images.[15] This was not considered an incremental improvement in the computer vision community rather a breakthrough that caused a rapid adoption of the methodology by others in the same community as well as in many others. Subsequent achievements using similar techniques for other AI tasks like playing games that require strategy and planning sophistication are mentioned next.

---

[12] The system by Google Deepmind that plays the ancient game of GO and has beaten the world champion in GO.

[13] "What Do People — Not Techies, Not Companies — Think About Artificial Intelligence?" HBR.

[14] Krizhevsky et al. "Imagenet classification with deep convolutional neural networks".

[15] Wherein, objects may belong to any of roughly 20,000 different categories such as dogs or balloons - see ImageNet.

These applications are all very impressive not because it was previously deemed that AI could not overcome their challenges, but that it was anticipated that reaching human-level performance in them would not happen anytime soon. For instance, regarding the game of GO people generally had the feeling that it's still a few decades away until AlphaGO beat the world champion in March 2016.[16] This was chosen by the Science magazine to be among the top scientific breakthroughs of the year.

However, all these celebrated AI systems are called "narrow AI" systems. The reason is simple. The AI system that plays a game, for instance, is not the same as the one that could drive a car. Not only it isn't but we currently cannot make it such that it could be. This inability has encouraged creating the widely-used labels of "narrow AI" and "general AI".

"Narrow AI" refers to those systems built for a specific task, whereas "general AI" refers to those systems we wish we could build but we don't know how yet; systems that could perform tasks in many different domains. For instance, a system that drives itself to a game studio, plays the game, also describes in natural language what the game is about would be an example of general AI. Why can't we make it if we can already build systems for each specific task separately? Of course, we could in principle just put a few systems together as separate modules. However, they'd have nothing to do with each other in the sense that they are different AIs and there is nothing shared between their experiences and data. That means that for every specific application you have to build the system from scratch. Making one successful doesn't help with the other at all and that seems too "narrow" as compared to the intelligence that humans seem to possess. In contrast, the label "General intelligence" was introduced to roughly target and chase human-like intelligence.

There are significant issues with all these labels like general AI, human-level AI, human-like AI, true AI, etc. We'll later object to all these labels, including what I am about to use next!

Almost everyone considers the mission of AI to reach some kind of general intelligence or for "solving intelligence" to mean creating an "Artificial General Intelligence" (AGI) system.

AGI is a very popular term and widely used to refer to a system that is as general as a human being in the range of tasks it can handle or problems that it can learn to solve. Researchers widely

---

[16] "Feeling" the key word in this statement, which tells us we don't have a guiding theory to tell us how easy or difficult tasks are and the reason for that should become very clear in this book.

consider marching towards AGI to be equivalent to marching towards more intelligent AI. That's an assumption that may not be true, in fact we'll eventually argue that it isn't necessarily.[17]

Regardless of what intelligence could best be represented as, AGI would likely still remain a target for AI research. A non-controversial reason for that is as follows. Consider applying AI to solve tasks that may vary stocastically[18] from time to time but only slightly. Even a slight variation for a narrow AI would mean a new task. Having to build or train the system from scratch is obviously inefficient and not scalable in this setting.[19] A controversial reason is that there are tasks that reaching human-level performance in them requires some combination of skills. Each skill could be acquired by a narrow AI system whereas the combination couldn't. An example is an AI tasked with teaching a human how to drive like a licensed trainer would. That would require some "intuitive expertise" in several domains: linguistics, psychology, physics, and so on. Such tasks or problems that require an AGI level competence, are sometimes called "AI-hard" or "AI-complete" problems. There is no consensus as to which tasks are AI-complete/hard and which aren't, hence the controversy. Conversely, some of these tasks are used as a test to tell when a system can be granted the AGI degree. An example of such a test is put forth by Steve Wozniak, co-founder of Apple, called the Coffee test. The test is for a system to walk into a random house for the first time and proceed to make and serve coffee and nothing else! Just like there isn't any consensus for what is considered an AI-complete problem, these tests too are all questionable as to whether any of them would actually be the right AGI detector. Yet, there are other widely used labels or categorizations too, such as "weak AI" and "strong AI". These are even more problematic and we won't entertain them directly.

Certainly, current AI systems, as impressive as some of them may seem, are all quite "narrow". So if AGI is the real deal and it isn't solved yet, why are we experiencing this giant hype around AI then? What's beneath all the excitement?

---

[17] At least in biological learning this goes back to Seligman's "Preparedness to Learn", 1972. Arguing that there is no general ability to learn, but that each species is more or less able to learn what they are supposed to relative to their "instinctive wiring".

[18] Roughly speaking, stocastic is a broader concept than random and does not require statistical probabilities to exist or be defined, the way a random variable does.

[19] There are research thrusts and practices such as "transfer learning" that could address a tiny fraction of all aspects of this problem but they don't constitute anything like an AGI.

The World of AI?

Consider some of these impressive systems like Google's current translator, a self-driving car's object recognition system or Alexa's voice recognition system — all examples of narrow AI. These systems have a lot of shared elements in them when it comes to their abstract algorithms[20], which are commonly referred to by the insanely popular phrase, "deep learning"[21], the title of the 2015 Nature article by the three pioneers of the technique: Yann LeCun, Yoshua Bengio, and Jeffrey Hinton.[22] So the generality in technique and method is one source of excitement.

Another source is the performance that they have exhibited on many benchmarks. They achieve higher accuracy in many machine learning problems without using any domain expert knowledge, wherein previously a lot of expert knowledge was being leveraged to improve the performance of the system such as in computer vision, voice recognition, or language translation. All that changed during the past decade. Deep Learning has become the state of the art. Google's products in all three of these domains are currently based on these competent deep learning techniques and are being used by billions of people.

The third source of excitement is how these systems scale with the amount of available data (number of distinct examples that the AI system can use to "learn" to perform better at a given task). These deep neural networks continue to improve their performance when given more data, even as other techniques start to plateau. So it feels like regardless of the task, you could always improve AI's performance by making the system bigger or giving it more data. That seems pretty general, right? Almost everyone jumps to say yes. Hence the excitement and billion-dollar investments in companies who'd say they'd do just that!

If many people who are too far from AI development think it's almost already solved, many of those who are closer to it think that it's at least on the right track to get solved. Although most acknowledge the existence of many shortcomings with current methods, their belief seems to be that deep learning possesses the right attributes (especially the three I just mentioned) to be a candidate for AGI and therefore we are on track to solve the remaining technical challenges on the road to AGI. An impression that certainly paves the way toward an over-hyped environment.

---

[20] Most commonly "neural networks optimized using a specific algorithm called backpropagation".

[21] It is not deep nor learning in any common sense of the word, only in a very narrow technical ML sense. In chapter 6, we'll discuss deep learning in the required depth.

[22] It's cited 17k times in just 4 years and Microsoft's semantic search for Nature ranks this paper as the top paper associated with Nature: https://academic.microsoft.com/search?q=nature

But before we get to the over-hyped aspects of AI, let me mention the last and what should be the most important source of excitement. That is the availability of large amounts of data and compute power which in turn resulted in the emergence of the exciting fields of "big data" and "data science" earlier in the decade. Data science as an independent discipline is still being shaped and in its infancy but it's much broader than what currently AI stands for — see chapter 8. Data science exists with or without AI. The reverse isn't true.

The main algorithms in deep learning are quite old, going back to the 50s and 80s. Except, in the past decade we have had much larger amounts of data and compute available to throw at those algorithms, which did give rise to, you guessed it, the deep learning revolution!

As I mentioned, these techniques are already commercialized at scale. Most maturely within the IT and internet industry, but certainly not restricted to them. Expansion into other industries has been happening, though still in its infancy. One way or another, deep learning and machine learning, in general, will be transforming many industries.

What do we know for sure about the next 10 years? People will continue trying to apply today's known flagship methods (such as supervised learning[23]) to almost anything. It doesn't mean all these applications will be successful but the number of low-hanging fruits is surprisingly large. It is very hard to overstate the many valuable problems that can be cast as a simple machine learning problem awaiting promising solutions. Let's consider some of the less talked about examples. Innovation in manufacturing is notoriously more difficult in which adoption of new tech is proven to be almost always slower than in other industries. Can "learning" affect that in any way?

Let me give you an example in device manufacturing. One can produce low-quality devices missing many features and then correct them with machine learning. Instead of letting the hardware do the job, you compensate with the software, which has to be learning from data so that it can correct for the shortcomings of the hardware and add new features to it as well. An interesting case maybe with cameras. Nowadays when you go buy your smartphone, you may care a lot about how many cameras it has on the front, the back, with how many megapixels each, what the quality of the lens is, etc. In the future even with a crappy lens and low-resolution camera you may get the same quality with more effects resembling the kinds of things that your eyes can do because the software is going to correct for limitations of the hardware just like your brain's vision system does for your eye. Your brain does a lot to cover up for the limitations of your eye, which simply gathers extremely

---

[23] We'll cover all kinds of machine learning later.

puzzling raw signals[24]. Deep learning methods have shown a lot of competence in interpreting these raw signals, that is going from pixel intensities to recognizing objects or even generating new ones useful in fancy image and video filters. Think of those popular filters in popular social apps like Instagram or Snapchat. Similarly, the same methods have tremendous potential in changing the device manufacturing landscape, somewhat beyond the current imagination.

Having said that, deep Learning being the state of the art, is not without its flaws and disadvantages. Throwing deep learning methods at every problem isn't the right approach. The industrial domain is full of highly complex and non-stationary[25] use cases that don't quite lend themselves to current deep learning methods. In these cases, a lot more data science and domain expertise are needed to come up with more efficient and appropriate approaches. However, such expertise is hard to come by. On the other hand, there are nowadays lots of freely available tools and training for almost anyone with any background who wants to learn deep learning and use it. That goes along with a very large and fast-growing community around it. There are now students graduating from medical school with full literacy in deep learning. The same goes for agriculture and many other fields too. Practitioners and domain experts are also being widely trained to be able to apply cutting-edge deep learning techniques and tools to their respective projects.

The sheer human-resource scale factor will be driving massive adoption of deep learning in every domain for at least a decade to come. The fact that deep learning may not be the right tool in many settings would not matter. It's the tool that almost anyone would have and know how to apply even if they wouldn't know the full implications and missed opportunities in doing so. Methods that win the adoption game, never have to be the most efficient ones. Just like how products that win over markets, have barely anything to do with providing the best solution possible. All these are well-documented facts. The winning factor is convenience, some combination of easy and quick.

Bottom line, these methods will be applied to many transformative use cases in an agile fashion and if they don't fully work, more data science and engineering will come to rescue the day and iterate to make them work eventually. As a result, lots of things in our world will look quite different or rather work quite differently in 10 years' time.

---

[24] For a dazzling articulation of this show of mastery by our brain, I highly recommend the book titled "How the brain works" by Steven Pinker.

[25] That is where the probability distribution of data or the target phenomena keeps changing.

As it stands now, there is no need for any breakthroughs in AI to facilitate this transformation[26]. That is again not because AI is solved or is getting solved but that there are astonishingly many problems and use cases that we can do something valuable about with a little bit of clever machine learning. This massive economic potential is the driving force, which is currently put mainly behind deep-learning-based technologies, giving it the momentum they have. This settles why there is a lot of hype and excitement around AI and why some people call it a revolution.

This brings us to the rest of the narrative that the public keeps getting fed. First you're told that what we currently have is "narrow AI" with some promising recent breakthroughs. Then the story that is often referred to as a technological singularity, goes as follows.

We will at some point, if not soon, reach AGI, or human-level intelligence. Progress will continue because it cannot possibly stop unless we are all wiped out including the AGIs. That will result in a super-intelligence, an intelligence that is superior, at any conceivable task, to any human that has ever existed. Superintelligence can program itself and improve. The more intelligent, the more it will know how to become more intelligent. Therefore, an intelligence explosion will occur because the intelligence-improving iteration speed of a super-intelligence is very short compared to our time scales e.g. our software release cycles.

You can already feel the horror embedded in this sticky story. Almost everything we care about around us including ourselves came to us as a result of human intelligence. Now everything could be at stake. That's why many philosophers find this such a hot topic to work on. Nick Bostrom, the author of the popular book "super-intelligence" and among the original promoters of this narrative, calls it "philosophy with a deadline", regarding what this phenomenon demands.

It's very appealing. Progress is not going to stop. Given that our intelligence is not infinite, it passes us at some point. So it could overtake us! Seems like bulletproof logic, right? Not so fast! This story is not even wrong! It's utterly meaningless. The only reason so many people believe it and it's a dominant narrative is because apart from its straightforward logic, it leverages a lot of our powerful cognitive biases.

Let me briefly mention what these biases are without applying them directly to the details of the story:

---

[26] On the other hand, breakthroughs in data science are needed. We'll get to that later but data is for sure an overlooked part of the equation.

1. It's a simple story and we get really intimate with simple things. That is because we can absorb them with less energy and also utilize them with less energy. This is a form of survivor's bias in the evolutionary sense, not to be confused with Occam's razor. We tend to believe all simple things that seem logical over more twisted things. Especially if the twist requires us to change any of our preconceived notions or beliefs, which brings us to the second cognitive bias.

2. That is the well-known confirmation bias, we tend to be biased towards selecting out the evidence or explanations that confirm our own assumptions and beliefs.

3. The first two biases are tiny compared to the effect of the third bias, that is when our primal fear is invoked. In which case, we tend to be over-welcoming a lot of false positives. Our bias is that we think it's worth it. Even if the fear is irrational, we tend to get blind-sided by it and our brain somehow jumps over examining the claims and arguments properly. That's why maintaining intellectual honesty is often difficult. It does require one to really slow down and say no to many quick opportunities to crank up fame. Intellectuals who keep echoing such narratives from different angles are themselves victims of their own biases, many of whom are respectable and prominent professionals. We'll see what we can do in this book by "slowing down" and being much more careful intellectually.

   Appealing to people's sense of fear is highly effective to further your agenda. It's no surprise that there are many popular TED talks echoing these common narratives around AI. That brings us to the fourth bias.

4. The "repetition or attentional bias". If you see or hear something like a self-consistent argument, many times by many people, you tend to believe it more than if you heard it only once or you just thought of it once yourself. The effect of hearing it from a prominent person or a public intellectual is technically a separate bias but I am bundling them together.

These biases are precisely why this story has stuck around as predominant and prevents people from grasping the true naivety of the story. And this is not the only misleading story around AI appealing to these biases, there are many others floating around. I am listing the biases here simply because it's good to have them in mind while we listen to arguments by big names.

There is no question that AI will continue to get very powerful (we'll clarify what powerful means) and we must work on many safety issues in that regard. It's also true that solving AI safety problems can be more challenging than just mindlessly creating some powerful AI. Some of them definitely are, simply because we have to factor in humanities and social science, which unfortunately we know too little about. All such fields with the concept of humans as the central ingredient, are not mature sciences yet in comparison to physics or established engineering disciplines.

However, this challenge in understanding humans is also why people have been mistakenly separating AI from AI-safety when in fact such separation doesn't quite make sense. Stuart Russell, a highly renowned AI professor at the University of California at Berkeley articulates this very well: "you can't separate the two just like you can't separate building bridges from building safe bridges. They are deeply integrated under the discipline of civil engineering." Thanks to his efforts, as well as those of Future of Life Institute (FLI) and others alike, there is a newly-forming vibrant field, mainly referred to as "AI safety and beneficial AI", dedicated to such challenges.

It's true that AI safety and similar fields that don't yet exist, must be there and actively mingled with AI engineering, but that's true regardless of whether AGI or superintelligence are meaningful things to talk about or not. As you may have guessed, we will argue against the meaning of such terms. But we don't want to just criticize the terms, we want to figure out what we should replace them with in this book.

So before we talk about those fancy terms, let's start with AI itself. Even if it's your first time reading about AI (quite unlikely), I have repeated the acronym so many times here that you kind of feel it's something real. It has objectness[27] to it. My repetition past one or two pages makes you numb to questioning whether the term makes any sense or not, that is me tapping into your "repetition bias". The same thing happened with the people in the industry and the business world, it's just a tag on a bucket that way too many things get dumped in it.

Suppose you're a skeptical soul who wants to start from scratch with AI. Well, if you start questioning everything, you'd get stuck right off the bat with the definitions of things: What is artificial? What is intelligence? You immediately see that by those we mean very superficial things like: artificial means non-biological, intelligence means this thing that humans got more of than animals, etc. We don't want to get into definitions and prior work here, though we'll do some of that later. For now, it suffices to say: there are no satisfying nor useful definitions of intelligence let alone any theory for it.

---

[27] An innate property that our cognitive machinery is born with, and we'll talk more about it in chapter 3.

In fact, most rigorous treatment goes back to one of the main figures behind the birth of computer science and AI, Alan Turing, who basically said: you know what? Let's not define intelligence because we don't need to! That gets us nowhere other than endless debates with philosophers. Let's instead have a test that says what or who is intelligent, i.e. the well-known Turing test. This has literally set the course and research attitude for everything in AI till now.

I'd like to ask you: what if he was wrong or rather, extremely incomplete relative to what we need now? Many would tell you that the original proposal has many issues associated with it, but I am not talking about the choice of test or variations on it, that's a completely different topic and still considered within the Turing world. I am encouraging a step outside of that world by simply asking: What if we take defining intelligence seriously? If you also have any curiosity itch for that and would like to go deep into it, you're reading the right book. My goal is to convince you that if we did, it would instantly affect our work in AI and many other disciplines. The fictional story at the end of this volume describes a scenario where that becomes a must.

AI as a Label

I'd like to draw your attention to the fact that we have all these fancy AI-based terms way before intelligence itself is well-defined. Way before we've had a proper chance to discuss if or how defining intelligence and setting more robust foundations for it could change things. It can be seen as absurd, putting the cart before the horse, the fact that most people are so willing to jump into talking about "possible paths to AGI", making it perhaps the second most controversial topic next to what consciousness is and whether AGI or AI should be conscious or not, before working out any satisfying fundamentals for the concept of intelligence.

Well-defined or not, AI is a term that is stuck and is not going to go away. But it is not even a good label necessarily for the data-driven technologies and products it may represent and here are some basic reasons why:

1. The first instinct of most people unfamiliar with AI, is to compare AI to themselves. A natural sense that says "okay, I am intelligent, so it's like me, except artificial. What if it gets too smart, oh shoot, let's at least make some money with it or integrate/hybridize with it or something...". It just sends the wrong message.

2. The wrong and illusion-based fear of the rise of potentially evil and embodied intelligent agents e.g., robot apocalypse, keeps us from paying more attention to real challenges facing

our society, all due to increasingly more capable software technologies, such as potential AI bias problems in powerful recommender systems.

3. It leaves out the intelligence of so many other things. Other examples of intelligence that we ought to focus on technologically and yet most people may not regard as intelligence, an example being many components in an intelligent infrastructure that have no human or single-agent analog for. Think of maps for self-driving cars. We, humans, use Google maps or Apple maps or what have you, which are the same for all of us, because we all look at it, interpret it, and use it the same way. That's not the case for autonomous cars. There may exist (not that it should) many different unique digital maps based on different technologies and for self-driving cars to communicate and to dynamically and collaboratively enhance their respective maps. Think of the critical collaborative intelligence from the car's perception level to smart city level coordination. This is an example of an intelligence that has to exist at the network level, infrastructure level, and the level of the content being exchanged. That's a very different kind of intelligent system compared to a single AI agent playing games with people.

4. Because it's not a well-defined term, it leaves room for anything a little bit smart to be called AI. There are no judges in the industry except for customers and as for that, people are just confused. In the business world, people's knowledge typically boils down to a Venn diagram that states: Deep Learning is a kind of ML, ML is a type of AI and AI can be more general, for instance, a rule-based system that can act "intelligently". What is actually intelligent, is left out to interpretation and this gap can be misused massively and it has been.

Now, good or bad, hyping up any label has consequences, both positive and negative. First, we should acknowledge that an enormous amount of positives and progress has come out of the AI trend and somewhat because of the hype. However, given that the positive effects have gotten most of the coverage in the media, it'd be more useful for us to stay mindful of some of the negative effects of overhyping. So just to balance the playing field a bit let me mention at least a few of those:

●   Almost anyone who is working with data or building data-based systems is effectively forced to call what they are working on, "AI", as it's currently hard to "sell it" under any other label. Even in academic journals, the usage of the term is hardly tamed.

- Due to the hyped-up demand, the focus goes on training people fast and at scale, instead of at depth.

- Without depth, misconceptions continue to grow and the hype continues to be re-fueled mostly by the startups whose funding and survival depend on the hype.

- Built upon the hype, the public's confusions can likely grow. Confusions due to a general lack of deep knowledge and training, leading people to take away the wrong messages from PR releases.

- The over-hyped environment pushes most towards over-promising and under-delivering, which could lead to another AI winter. Though should it ever happen, it'd be a very different winter, simply because there are many products already out in the world which need continued support and improvement.

Venture capitals and major institutions have already stopped buying too much into the hype. Setting aside a few billion-dollar investments into some ambitious research plans, starting in 2018 there has been more focus on actual business plans and a major shift away from trust-based or hype-based financing. All the while, the statistic of CEOs mentioning the term "AI" on earning's calls has been off the charts since 2017 relative to the case for other tech buzzwords in the past.[28] You almost universally hear phrases like "we are already using AI to do X and improving Y by Z percent" or "with the help of AI we can do X or Y" and all our attention goes to all sorts of really cool applications addressing various pain points.

We should keep in mind that while CEOs and the industry may drop any label at any time in chase of another, legal documents, minds of people, movies, etc. may not. Unfortunately, the same misconceptions and bad terminologies in the field of AI, have found their way directly into congressional bills such as "Future of Artificial Intelligence Act"[29] and Trump's AI act. The laws and policies are being enacted around these definitions with direct consequences for tax-payers and are prevalent among many countries now.

Having said that, let me also express my sympathetic view. The Speed of development and coverage of the AI trend has been very fast. When too much noise is dropped on anyone's radar too fast, the first act of management is to put a single umbrella around it all with a single and simple tag.

---

[28] https://www.cbinsights.com/research/artificial-intelligence-earnings-calls/

[29] https://www.congress.gov/bill/115th-congress/house-bill/4625/text?q=%7B%22search%22%3A%5B%22the%22%5D%7D

Our reaction is governed by the speed of noise. Past a certain threshold, we just work with a sticky catch-all label.

So, our task in this book is as follows. Given that AI is the terminology we have to live with, let's at least give it some real meaning. That is to understand what can be fundamentally meaningful about it and what cannot. The right definitions are deeply connected to understanding which can speed up progress. Here are a few reasons why having good definitions and establishing a solid foundation actually matters:

1. Lack of them results in many wasteful debates and arguments, as it already has.
2. Wrong definitions can mislead actual work and the communication of the work.
3. Right fundamental definitions can lead you in the right direction, and bad definitions cause missed opportunities.
4. Wrong foundations can affect the influx of talent to the field in a negative way by giving away the wrong impressions.
5. Long after we've set bad definitions based on questionable assumptions, we likely forget many of those assumptions. However, once things get too complicated due to any of those wrong assumptions, we have effectively lost the ability to find the culprit. We let it escape us trivially.
6. Investors and collaborators (across other fields for interdisciplinary work) would be better positioned to help and support our efforts. The public can also follow us better with better management of expectations.
7. The right foundations can help set matters of safety, policy, etc. on a more natural path of least resistance.

What About the Brain's Intelligence?

So far we've only discussed that we have to define intelligence and set the foundations right. What about the human brain? After all, it exhibits the best example of intelligence. Or does it? As natural as it may be, let's not make that assumption quite yet. What's true is that the brain is the best studiable example of intelligence we know of. The brain can indeed tell us a lot about intelligence and it already has.

Both fields of neuroscience and computer science were being developed as scientific disciplines concurrently in the 1940s and 50s. By the 60s, they were independent freestanding disciplines. Cognitive science and AI were born in the same timeframe as well. Obviously they all have something to say about intelligence and when it comes to our understanding of intelligence there has been a lot of exchanges between these fields assisting one another in various forms. We will dig a little bit deeper into that in chapter 3, as no talk of intelligence can be considered complete without mentioning the brain and the mind.

The most sincere writing I know of to date purely dedicated to intelligence is the popular book by Jeff Hawkins titled "On Intelligence". Frustrated by the lack of any books on the brain's intelligence with a theoretical framework to understand and potentially replicate it, he wrote one. Many have found Hawkins' book quite stimulating. For instance, Andrew NG, a prominent AI figure and a pioneer of the Google brain project, has said that reading "On intelligence" had a big influence on his mindset about AI.

Regarding his approach, Hawkins writes:

"... I refused to study the problem of intelligence as others have before me. I believe the best way to solve this problem is to use the detailed biology of the brain as a constraint and as a guide, yet think about intelligence as a computational problem — a position somewhere between biology and computer science. Many biologists tend to reject or ignore the idea of thinking of the brain in computational terms, and computer scientists often don't believe they have anything to learn from biology."

He states that he believes we cannot build truly intelligent machines before we understand how the brain thinks and he has dedicated the rest of his career to "brain science". The guiding belief in his neuroscience institute is that intelligence is in the brain's neocortex and the best path to building intelligent machines is to reverse-engineer the neocortex.

I subscribe to two very crucial aspects of Jeff's arguments. One, as he mentions in the quoted text above, is the need for an interdisciplinary approach towards understanding intelligence. And the second is the need for a rigorous theoretical framework that can guide experiments. Both are very near and dear to what we are going to advocate here, except based on a very different philosophy.

When it comes to the interdisciplinary approach, the positioning we'd follow places intelligence closer to be eventually a topic of physics, rather than biology. Though we won't get to talk much about that before chapter 9. As for a theoretical framework, I am by no means going after a neuroscientific one. One reason is that I do not believe that reverse-engineering the brain as fruitful as it may sound will by itself lead to a fundamental theory of intelligence.

Cracking the codes of our cortex, as an extraordinary giant of an accomplishment as it would be, won't give us all the answers as to why they are that way and not any other imaginable way. Those "why" questions can only be satisfactorily answered by a theory that is completely brain-independent, such that the brain would be only an example of. The problem is that we forget that the brain is just an example of intelligence! In this sentence, our focus almost always goes on the words "brain is" rather than on the words "an example".

The silent assumption seems to be that intelligence can be defined by what the human brain has, and that if you understand the human brain you've understood intelligence. This is exactly the assumption we are going to break. Moreover, we argue that having a fundamental brain-independent theory of intelligence may not only be very helpful in understanding the brain, but a necessary ingredient at some point. Therefore, a central thesis in this book is

To fully understand the brain's intelligence, we must go beyond the brain, abstracting away the differences between different examples of intelligence.

Just like we argued in the last section that AI needs a fundamental foundation, cognitive science also should have a deeper guiding theory to treat human intelligence truly as an example. Without getting into the details that we cover in chapter 3, cognitive science since its foundation in the 50s was supposed to emerge as a freestanding discipline to understand the mind and intelligence by drawing necessary elements from various fields, including AI. That hasn't quite worked out and nowadays the field is dominated by psychology and most AI researchers find themselves in a different camp.

It's helpful to think of AI researchers as to be roughly belonging to two camps:

1. Mostly dominated by computer scientists. Methods are empirical and whatever works, rules. The dominant philosophy is that we may draw inspiration from the brain but for the most part who cares if the brain does it differently and humans aren't that good at many tasks anyway.

2. Brain-related camp. Drawing direct guidelines and lessons from psychology and neuroscience. The dominant philosophy being that we should work on building human-like intelligence (by studying the human brain at least in part).

There are endless debates between these camps with arguments like "my algorithm is better than yours", or "Ok. but the brain does this thing differently, which you can't do with your algorithm". Currently, these fields are peers. I believe that the views presented in this book have the potential to set the stage for a unifying framework for different approaches towards intelligence. And that unifying roof is what I call "Fundamental Intelligence" (FI)!

This book is the only attempt to date to be agnostic to non-fundamental differences between cognitive science and AI. I believe in the physics way of doing things. That is, going the fundamental way, finding some first principles and reasoning up from there. Scientific psychology towards the end of the 19th century gained its momentum that way, being inspired by the style and quantitative accuracy of physics and trying to get to a similar level of understanding. The only mistake then was to imitate physics instead of borrowing the physics way of doing things. We'll get to talk more about that later. Here I'll just state that we are going to draw a lot of inspiration from physics and in the end argue for the necessity of establishing a new field one may call "*Intelligence Science*", which would be to Computer Science, what Physics is to Mathematics!

One may wonder why no one has done it yet? We've had so many smart people doing pioneering work in AI. Have they all missed it? Well, it's not that shocking when you learn about the nature of the philosophies they were building upon. More on that when we cover the philosophy of cognitive science.

Opportunities That Require a Shift in Attitude!

As we alluded to, our goal is to shift some focus away from the fancy and sometimes intimidating labels, and paint a different big picture and raise truly foundational issues.

When it comes to intelligence, ideas naturally vary among different communities of researchers. Nevertheless, everyone would certainly appreciate a more universal understanding of the big picture. Not just one that has a much wider and robust consensus behind it, but also one that is about how things should be, rather than how they currently are. The obvious question is: what's required to get there?

We collectively know so much already. The challenge is that a lot of this collective knowledge is spread across various fields and domains. That reduces obtaining a big picture to be just a matter of proper integration, not to suggest that it'd be a trivial integration.

So, let's talk about the concept of proper integration for a second. My observation is that the single key enabler of every innovation, scientific discovery, technological leap, or game-changer business idea, is always the same: putting two key ideas/components/domains together from previously disparate origins or contexts, such that you do no work on improving upon any one of the components alone. Your work would be strictly limited to figure out how they may go together or *how to put one in the context of the other*. Let me give you a few examples:

1. Business example — Uber: Cars, drivers, and passengers were there, and so were smartphones, applications, and all the infrastructure behind them. One had to put one in the context of the other. So uber essentially made the true asset to be the smart application and not the physical cars.

2. Technology example — Search engines: Internet and a lot of organic content production were already out there, databases, various indexing, and querying techniques were also there for decades. So not only the need for search engines was obvious but also all the elements were there such that one just had to figure out how to put them together well and obviously Google did a better job by laser focusing on the integration problem (at least initially).

3. Science example — Hawking radiation: Classical blackhole horizon was there, Quantum mechanics was also there. Stephan Hawking asked: what would happen if I put a bit of Quantum mechanics in this otherwise classical context, and out comes prediction of the Hawking radiation and subsequently the exploding field of black hole thermodynamics.

I could give you many more examples in any context, from the invention of statistical economic theories, Shannon's information theory to a business like GOPro. In fact, I have never been able to find a counterexample on any idea that ever took off in any domain. This is another way of phrasing the fact that *timing* matters more than any other factor in the success of startups, innovations, and discoveries, etc. However, there is no need to feel victimized because timing is out of our control. It's true that it's pointless to shoot for good timing but you can always examine whether you're merely integrating or doing grounding research on an ingredient. The latter is your simple gauge to detect bad timing on trying to integrate.

Alright. Even if integration was a clear matter, it would still be a tall order in the case of a bigger picture for intelligence as it requires both breadth and depth, i.e. simultaneous specialization in multiple fields.

Can we just facilitate more collaborations among different disciplines? We could, however, any such collaboration is usually structured toward specific applications based on the core agenda of only one of the parties involved. The communication overhead across different disciplines is high enough not to allow for an independent inter-discipline to emerge with its own independent depth and philosophy. That seems to have happened to "cognitive science", where more than half a century of trying, still did not result in the emergence of a new independent field as many had hoped. We can learn a lot from that effort.

Having said that, I believe with a shift in perspective, the timing is absolutely right for a field like "intelligence science" to be born. Sometimes in order to make progress, a different way of looking at things is all you need. Therefore our attempt here is to offer new perspectives with which to look at intelligence. Let me give you a sense of the approach.

What nature has taught us is that things should be generally simple. Even the most complex things are simple after you understand them or have the right theory for them.[30] This links simplicity to understanding. But simplicity is a controversial topic at least in physics. What about understanding? Clarity on the concept of understanding is undeniably useful for the field of AI. For instance, in Natural Language Processing (NLP)[31] and Natural Language Understanding (NLU), where understanding is just an aspiration without a fundamental or clear definition.

We are always told not to reinvent the wheel. But becoming a physicist demands quite the opposite. You must reinvent/rediscover many things for yourself or else you won't get far. Going through that process has given me some understanding of what understanding is:

If knowledge is knowing what something is,
understanding is knowing why it isn't anything else!

We don't have any universally quantifiable degrees for understanding yet, but at some point we could, directly based on the above intuition. This intuition alone may be immediately applied to many disparate contexts. In representation learning of words, for instance, it may natively drive you towards something like "negative sampling"[32] i.e. feeding the wrong context-word pairs to the

---

[30] Planet's orbits are elliptical, right? One ellipse per planet. That's what Kepler discovered. However, before that, people kept adding 100s of circles on top of circles, called epicycles, to fit the data. The point to keep in mind is that complication is always easier than simplification!

[31] No need to memorize these acronyms. They are just widely used terms in AI, in case you are interested in researching them further.

[32] A technique in representation learning.

system, so that it learns *what the word ISN'T*. However, that is still far from "WHY it isn't". In order for the system to get to that "why", it should be able to imagine other plausible scenarios. And for that, you need a mental model of the world and so on. Only then, we'll get to quantify understanding relatively easily, based on how non-plausible the imagined scenarios were, how many other scenarios it came up with, how big and rich was that self-constructed world, how self-consistent was it, etc. Similarly, you may argue that some level of understanding of this kind is also needed to truly handle the data veracity problems.[33]

Bottom line, to understand, you must ask what else it could be, instead of what it is or was. This should suffice to convince you that the dominant AI technologies that is solely built on a set of correct and well-cleaned examples will never get to *understand* anything!

The point is that without fundamental principles we don't know what it means to march forward efficiently. A few benchmarks (and error rates against them) barely locate us on the land of intelligence and in many cases, may give a false sense of true progress where there is none.

Taking how the system works (specifically how efficiently) into account rather than simply what the system accomplishes, can have huge implications for the world of research. That is, to introduce more basic research into research. Everyone agrees that we need more benchmarks and that's exactly the course of progress. The issue we raise here is with the current lack of any benchmarks for benchmarks! How could we measure the relative quality and meaningfulness of benchmarks?

Obviously, if we don't know how to measure something yet, it doesn't mean it's not measurable. If we could have theories about the "how" of intelligence, it could shake the current realm of benchmarks. Let's advocate for at least some group of researchers to work on building rigorous foundations to guide that process, starting from scratch. That requires advocacy because when pursuing a radically different approach you may have a lot more ground to cover before you can measure yourself against established benchmarks designed by established techniques. This has already proven to be the case for artificial neural networks. Before the deep learning revolution in computer vision, those who were hand-engineering features could pass almost all established benchmarks better, and there was little to no proof that neural networks were going to dominate

---

[33] Veracity refers to challenges in data-based systems where part of the collected data that is supposed/assumed to be correct to a good approximation, isn't actually so due to various biases, lack of credibility/truthfulness/objectivity, etc. Many consider veracity the most important and tricky challenge in big data, certainly relative to high volume/velocity/variety/ other big Vs of big data.

that field, and logically not a lot of people wanted to work on them. Now it's obvious that hand engineering features was never going to go far. Could we have possibly seen it earlier?

The lesson to draw is that how we do things, particularly their efficiency, matters!

Sounds Like Philosophy?

We do need to be concerned with philosophy, at least where there are reasons to believe it's going to strongly affect the foundations of AI and the course of scientific progress in the future. All sciences were once strictly in the domain of philosophy. There may exist some border that separates what portion of philosophy would fall into the lap of science in the future and what will remain only a matter of philosophy. However, this boundary and its exact geometry are only visible to the eyes of time. To more pragmatic minds, this is precisely what justifies a simultaneous practice of both philosophy and science.

Even if you're purely interested in the progress of science, philosophy can be of crucial importance. Consider some given period in time, wherein different methods, theories or formulations are not differentiable scientifically. That is, they would all give or predict equivalent results for the kinds of tests we are performing or can perform today. Today's differences between them are only philosophical and psychological. They do indeed differ in their potential in facilitating further progress. They may give the researcher different ideas as to how to go forward next when there are no other guiding lights. Feynman used to summarize this point beautifully. His advice: "take the world from another point of view".

The problem starts with the scientists not doing their own philosophy work. I am aware that this statement throws a professional philosopher off his or her chair. But many good philosophers are also scientists at heart or science-trained. Everyone is following some set of philosophies whether they are aware of it or not. In physics, there is some silent resentment towards philosophy and for good historical reasons.[34] I suspect however that if physicists themselves had worked explicitly on appropriate philosophical frameworks around their own work they may have been able to feel more positive about philosophy after all.

I believe that at this time there is an extraordinary opportunity for many other sciences especially for computer scientists, statisticians, and software engineers to build or participate in

---

[34] See Steven Weinberg's "against philosophy" arguments in "Dreams of a final theory".

building stronger philosophies around their work, its meanings, and implications. To help pave the way to integrate better relationships with humanities and social sciences, both of which are currently under a violent influence by technology.

On the other hand, humanity's opportunities and their associated challenges, don't know or care about our superficial academic lines that divide up disciplines. Some giant missed opportunities lie in the currently empty space in between these territories. So I hope I can convince you that if we ignore these lines and focus on whatever it takes to define the problems better and subsequently address them, the prizes can be unparalleled. The same mentality allows me to risk stepping on the toes of various disciplines beyond my own expertise and in some cases massively oversimplifying what they represent.

That means beyond trying to redefine what intelligence is and what AI could mean, we should discuss in-depth why and what for, should we even want AI. You may say well, there is the obvious massive economic growth that it can lead to or you may bring in inevitability arguments to say it's a moot point to discuss. Those may be obvious and are already talked about in many other places. What's not talked much about is that economic growth on its own is mindless. We must work much harder to cross it properly with humanities. To me, the best thing about AI is that it's making us debate ourselves. Questions that philosophers around AI have been raising such as whose values should the AI adopt, forces us to better figure out who we are, question our morality naturally, and so on. How can we NOT be super excited about the new light that AI is uniquely offering us to understand ourselves better under?

Let's begin our journey. In the spirit of starting from scratch with intelligence, I'll first tell you how mine began in the next chapter and we'll build our way up from there.

Chapter 2

The Missing Manual — Rethinking thinking

Take a second and think of the machines you have been using in the last 24 hours. What did you use to get things done and move your day forward? Most likely, your car (or some form of transportation), your smartphone (or some form of a computer), or maybe only an Espresso machine if you're on an electronic detox while being tied to your desk to write a book on "Rethinking Artificial Intelligence"!

You must have used some machine, and it did require you to know how to work it. A machine that has a manual associated with it. Surprisingly, even a kitchen scale comes with a manual, not that you actually read it.

Unless you just woke up from a coma that miraculously told you to get up and instantly read this chapter, I am certain that you have used at least one machine that unfortunately doesn't come with any manual whatsoever. I am talking about your brain.

"Wait a second, my brain isn't a machine, what about my mind, consciousness, and so on? There are differences, right?", some may react this way. Others may say "well, it doesn't need a manual. We know how to use it, it comes from WITHIN. You just invoke your free will to use it and its gears start to turn. Otherwise it works autonomously."

The curious one may entertain: "Sure I can use my brain but how can I know if I am using it the "right" way? And how can I use it the "best" way if all its possible states aren't known?" Then there is the mindful person who may identify this part of our brain rushing to tell us that we already know how to use our brain "just because we can think…", as the same part that often "lies" to us by such superficial rationalizations.

I was just turning 12 when a deep realization hit me for the first time and dropped my jaw south of my neck. It's one of those rare facts that never stops bubbling up anytime you start to declutter your mind just a bit. Especially when you consider its consequences for yourself, everyone you know, and the world. That is,

The machine that we use the most is the machine we know how to use the least!

Feeling a profound vulnerability, questions naturally popped into my head. "How are we dealing with this fact? What is or has been people's strategy around it? What about myself?" asks the 12-year-old me. With looking around comes another alarming observation for him. How mindless everyone was about it as if it wasn't a big deal. "Maybe people learn from life experiences and write some ultimate effective manual for themselves by the time they're adults?" he wonders. But he was skeptical because assuming otherwise could help him explain more things around him and a lot better. Yet he couldn't fathom how anyone was moving forward with life mindless about minds without manuals. "Doesn't everything that happens or doesn't happen due to our existence, have to do with how we use our brains?" he is baffled. "Did someone or something reassure people that it's going to be OK? How can they even know what is OK? Even to be certain about that, requires them to at least have partial access to some form of a manual" shedding a lot of trust in all he thought he knew. "How do they even know what they're feeling and saying is what's really happening to them? Is what people tell me about what they think they're doing even remotely true? Don't they see how all their proofs are awfully circular?"

Growing up in an uneducated working-class family that was not religious in a country that has the name of one of the world's largest religions in its latest title, he was bound to learn about religion and the fact that the answers for most people come from it.

Everyone agrees that there are problems in the world of humans, as far as the well-being of humanity as a whole is concerned. What is not generally agreed upon is what those problems are, how hard they are to be solved and who is supposed to solve them, and so on. What about the largest root of all problems? Can people reach any agreement there even though we tend to just look out to blame someone or something as the root cause and end it there?

For the 12-year-old who was experiencing a very limited world at the time, these questions seemed inescapable. Little did he know the answer he was about to give, was going to set the course of the rest of his life. That answer was: "The root of all workable problems is that we collectively are not using our brains sufficiently well... relative lack of knowledge of someone traces back to poor thinking of some others, and so on".

This all came years before he had learned that there are places called libraries, universities, giant bookstores or that books on all topics do exist, let alone computers or the internet. So there was a lot he didn't have a clue about, but he did know what was the most important thing that he needed to know or go after. *The missing manual.* Yes, of our brains and minds.

That was the most important simply because that alone had the power to fix the simplest life plan that sounds universally optimal for anyone. Only two steps: "first learn how to use your brain, then figure out what to use it for and what not to." Quite similar to other versions of two-step life plans, such as "first find out the truth about the world and then try to change it a bit to make the world a 'better' place."

Easier said than done, indeed! He is stuck right off the bat: "what do I mean by using my brain? Am I just talking about 'thinking' without saying the word? What is thinking?"

Is there more to it than the common meaning like when people pause and say 'let me think'? It makes sense to have a dedicated word for when we have to pause and intentionally consume dedicated mental energy. But that's just a form of conscious analysis. It barely scratches the surface. What about how you direct your perception, attention, awareness, managing senses, interpretations, memory, beliefs, etc., and even emotions... they are all happening in your brain, what are those if they aren't part of thinking?

It doesn't stop there either. For instance, we know that what we think about or tell ourselves today could affect our subconscious thinking tomorrow and vice versa. So without even touching the massively controversial topic of consciousness, we know that its existence is only going to make defining thinking more challenging instead of lending us a simple drawing line.

But why bother defining thinking at all? We have a socially practical meaning for it, why is that not good enough? Well, it depends on your confidence in the common practices and the society around you. Struggling with these, it was clear to the 12-year-old that common understandings were far from what he was looking for. Following the two-step plan, he wanted a definition that could encapsulate all human responsibilities. But drawing any boundary for human responsibility passes the ball to coming up with some moral theories and then back to the brain and philosophical subtleties in defining thinking and on it goes.

For the pragmatic boy who saw the meaning of everything at stake, latching on to never-ending spirals of philosophical games wasn't the most appealing route. Idealistically, he settles with a definition for his responsibilities that cuts out the most work: "Any mental activity that I could possibly be aware of or become aware of later, or that is caused by some earlier activity that I could have possibly been aware of."

The main utility of definitions is to simplify things in the sense that we don't have to talk about multiple things that may be harder to talk about when we can instead focus just on one. Same here, the main advantage of this definition of thinking is that there would be no need to talk explicitly about responsibilities anymore (and where their assignments come from), which he had found to be quite subjective. He now had an easy way to name the best status a human should aspire to be in, namely, a "full thinker".

This definition enabled him to have a simple explanation for all problems as well as a single recipe to solve them: Most problems are there because "people don't think enough" and the solution is "to think more, not less!"

By now, you may be feeling that this is all naive cause things are way more complex. In particular, we often get told to not think too hard or overthink something, that we often need to rest the thinking mind so that later we could think better or be more productive, etc. Note that we are trying to rethink thinking, not to use the word as we commonly do.

By commonly, I mean to include the currently most convenient scientific picture too — basically just the scientific version of the same popular definition or usage of the word. That is, thinking is just planning and doing except in imagination, "offline doing", in contrast to the "online doing" governed by the motor cortex in the brain. This sort of thinking is most captured by what the Psychologist Daniel Kahneman, calls *slow thinking*, or system 2, as opposed to intuition-based system 1, which is much faster and mostly subconscious. The names 1 and 2 are exactly there to avoid association with particular brain regions. Yet, defining it as "offline doing", tends to associate thinking with very specific regions of the brain, and separating it from all other functions such as acting, feeling, and perceiving.

These unknowns around association or disassociations with different parts of the brain should alone be sufficient for us and for now, to see why there is a lot of room to rethink thinking, beyond the high-level decision-making function. For instance, there is no reason not to include even a simple recall as an act of thinking. Let alone the discussions of voluntary and involuntary thoughts, the existence of control over thoughts which are much more complex and only partially scientific at the moment.

What we are presenting here is one path to see some of the complications that arise as soon as you want to get rid of some naivety around remedying the missing manual challenge. As we will discuss in the next chapter, thinking is still one of the most unscientific scientific-looking words in all of science. And the idealistic definition we gave earlier is no exception. For the 12-year-old boy, it was bound to cause many paradoxes that I still find worthy of contemplation.

Suppose that,

1.   There exists some fair universal value system relative to which we can objectively measure the outcome of a person's choices and effects on the total system of self and the outside universe.
2.   Define a measure of "thinking" to be equal to this hypothetical objective measure.
3.   Even though thinking, acting, and outcomes are all time-dependent phenomena, ignore all dynamic aspects for now.

Right away with recommending "think more", paradoxes start to appear. The 12-year-old observed that many people were thinking a lot, almost all the time according to themselves but the results were poor. Even worse, thinking longer may lead to what psychologists call analysis-paralysis, where you're not making progress in thinking and just going back and forth and only increasing your duration of thinking until you feel paralyzed in the head. That cannot be good for fulfilling responsibilities. So mere duration of thinking couldn't be it, i.e. thinking more is not thinking longer!

So then, is it the amount of thinking per some unit of time? That's paradoxically tricky too in many ways. Many ways because there are many tradeoffs. For instance, simply loading your brain more would backfire and you get less performance. Yet, speed of thinking had to be part of the equation because there are times you just have a fixed amount of time to do the right thinking, or else you'd effectively proceed randomly. But what kind of thinking and speed would matter in that objective measure? Regardless of the kind, the methods of thinking ought to be crucial to care about too per this outcome-based objective measure. For instance, if you happen to approach a problem from a relatively simpler angle, you can go much further with less time AND less speed.

Consider teachers in situations where they see a student fail at some analytical task despite having all necessary background knowledge. It's possible for the teacher to clearly see where the student is going wrong in their thinking route. But almost impossible to see why the brain of this fully knowledgeable student went on the branch that leads to nowhere to begin with. Most people discard such observations instantly asserting that "yea there may be a root cause but we'll never know, so let's just get more practice". These are because the story that almost everyone is told universally goes as:

1. Having all the necessary knowledge is not sufficient, you also need experience.
2. The amount of experience needed depends on how smart or intelligent the student is.
3. There are different kinds of attributes of being smart or intelligent. Someone is book smart, another is street-smart, one has analytical intelligence, another has emotional-intelligence, and so on.

So then obviously the recipe that people give you is to "just go get more experience in whatever you think is important and the amount of experience you need depends on how gifted you are in it". Given the assumption of a missing manual, don't these feel like gross illusions? I hope you can see that apart from the fact that certain kinds of knowledge are only attained through experience, everything else about these messages is extremely far from the best understanding we could possibly gain and utilize one day. At least the 12-year-old boy firmly believed so. Never believed it's fruitful to think in terms of "talent, IQ, etc." only in that there are ways to use our brains better. It was bothersome to him that he couldn't find a way to establish how to do better on this whole using the brain concept and the thinking definition.

Since the methods, speed and amount, all affect the hypothetical objective measure of thinking while any single one alone would be paradoxical, it was obvious that the recipe had to be at least some combination of "better, faster, and more" thinking. How could you make a practical recipe out of "THINK BETTER, FASTER, AND MORE"? The answer for him was that he couldn't, at least not without some deeper understanding. Not that he was not annoyingly pretending it's a useful recipe to recommend.

The light at the end of the tunnel came with some clues on the fact that there are relevant brain-usage differences among people. Differences that made him feel like we could one day quantify, clearly define what they mean, and prescribe how to go about them.

Here's one example that we will revisit later in the book. He observed he knew the answers to many analytical questions instantly without thinking. That made him conclude that he must have thought about them in the past before knowing the actual problems he was going to face. It had to be "thinking without a problem"! That was a concrete difference that he could pinpoint, noticing methods that seemed like what must have been written under some special "learn"-button in the missing manual! A special function that connects all connectable dots as soon as you receive them by generating and going through a cascade of questions that does the job of connecting.

At the time these were just feelings and scattered observations at best, rather than any deep or quantitative understanding. However, they were sufficient to make the 12-year-old believe there are more fundamental concepts behind it all. Concepts that would shine a light on what a proper combination of "think better, faster, more" could be and mean.

So far, we didn't get into many other problems and paradoxes that these idealistic definitions of thinking result in. For instance, basing responsibility on any possibility of awareness instead of actual awareness, or ignoring environmental and biological constraints, brings about many technical and philosophical complications that are way beyond our scope here.

However, the resolution for such complications does not help with the underlying challenges that this section is alluding to. That is, even if we get increasingly subtle in the definition and increase its complexity to avoid these issues, it still would be just kicking the can down the road! The reason is that useful definitions must decrease complexity, rather than increase it. When that doesn't happen, there must be something more fundamental at play that we could target, rather than monkeying around with definitions. The belief in this book is just that, which is why we are going after fundamental intelligence, hoping for it to provide a much better platform to understand thinking which would then not only apply to human brains but also much further beyond.

That was not how I thought of it back at 12, arriving at a dead-end in trying to come up with any robust definition. Still, I had to proceed with the first step of the two-step life plan, namely, learning how to use my own brain! Following the analysis of what "thinking better, faster, more" could possibly mean, I thought maybe thinking about more complex things could provide the right training.

What is considered a complex thing? Here's what I had roughly targeted: tasks or problems that had many moving parts, and a variety of relationships among those moving parts in a way that I couldn't trivially reduce it into a set of simpler subtasks with less complexity. This was a deliberate choice to avert any major dependence on the speed of thinking or duration of thinking. Instead, it was targeting an ability to wrap the mind around some irreducible complexity and somehow get to a point that it wouldn't feel complex anymore! That could be by looking at it differently or by figuring out what to pay attention to and what not to, etc. I didn't know better back then. I still recall the analogy I had in mind: "wrapping your head around something more complex is like lifting heavier with your brain muscle".

Where do you find such complexities? Thankfully up to that point I hadn't heard about science or philosophy. So, I had plenty of time with my "blank slate" to wander around and invent problems and theorize about them on my own. It wasn't until I went to high school where science was properly introduced. Within the first week of high school, in one class, I found out most of those complex things I had been pondering about to formalize, could be just called PHYSICS!

Imagine my excitement when my caring teacher told me that the subject has been around for centuries and there are places called universities in which people are still working on it. The week after, I found a university nearby, and a kind stranger who escorted me inside into their library. Within a year I had studied most of undergraduate physics, going from solving textbook problems to finding myself deep in superconductivity, a subject that I got to fully appreciate only a decade later in grad school.

I fell in love with physics and with a blink of an eye 15 years passed and I had totally forgotten about the missing manual. I engaged in almost every major field of theoretical physics. For a year I even went deep into the experimental world with quantum optics and some computational physics of accelerators and so on. I was simply driven by "the pleasure of finding things out" as Richard Feynman puts it.

However, my subconscious goal remained to expose myself to higher complexity, a higher diversity of complexity, and practicing the ability to transform such complexities into maximally possible simplicities. This happens to also describe what a theoretical physicist does. Over time, I discovered two particularly interesting things.

1. Physics had made me smarter. That's a simple fact with a simple explanation. A) nature is "smarter" than us, and B) to figure nature out more efficiently one has to learn to think more like she does. Given A and B, it's obvious that doing physics can actually make anyone smarter! It had indeed taught me how to "think better, faster, and more"!

2. Eventually, I had enough experience across the fields that I could see patterns that were not just about physics, rather about our brains and the physics of "thinking" if you will. In particular, I clearly saw that my brain was following the same mental methods

regardless of the subject or problem domain, as long as the problems fell in the same "class of complexity". That seemed obvious post-observation: "of course! it's not efficient for my brain to come up with different ways of thinking just because the domains are different".

These brought me to formalize a topic which I went on to call "subject-less thinking" and experimented with teaching it for a year to classes of mainly non-STEM students in exchange for a physics credit they needed as part of their overall undergraduate curriculum in Santa Fe College. We will revisit these important topics in Volume III when we are better staged to do so.

These observations awakened my attention to the missing manual problem and the only mission of the 12-year-old, namely, making himself and others "smarter". That's in fact how I've always articulated it to others; using the word "smarter" in place of the mouthful "thinking better, faster, and more". After building the foundations for a successful long-term career in theoretical physics, I realized that I had forgotten about what brought me into it all in the first place. It was this old mission and to go through with the first step of the two-step life plan, which at this point I felt I had accomplished. It was moving on to the next step that was the real challenge, whether to continue the conventional academic life or do something about the old mission.

While reflecting on that, one day I happened to have lunch with Peter Littlewood, Director of Argonne national lab at the time and a prominent physicist. Looking over the Gator's stadium in the University of Florida's campus, he was describing his path beyond a conventional career as a theorist, in which if you do well you get to publish something actually major once every couple of years and the cycle continues. That's certainly an admirable route and outcome but to him, it was more like surfing academic waves that he had done very well on. He wanted to "see something major happen".

Having surfed on one of those waves already, I realized that I would be going to mentally experience very similar periods of challenge, had I continued on a conventional path. Problems and scientific developments come and go, funding environments change, and for a career to survive successfully, one has to be flexible on what they choose to work on or continue to work on. On the other hand, these cycles, whatever problems they happen to represent, are quite satisfying intellectually which is what physicists and most other academics are seeking after all. And in fact, people adapt surprisingly well to these academic waves.

So far, rather obvious but there is one thing that is subtle to notice as non-trivial. Just like how surfing on water waves is invariant with respect to which beach the surfer is at, there is something invariant with respect to the specifics of academic cycles or problem domains. The way successful academics formulate problems and solve them is independent of the domain. And that is yet another manifestation of "subject-less thinking", i.e. *our winning methods of thinking are invariant with respect to the subjects they are being applied to*!

But life is finite and these cycles, as enjoyable as they can be, don't last forever. So if our goal is to maximize intellectual satisfaction and solving world problems, doesn't it make sense to work on the subject of "subject-less thinking"? if our best methods of "thinking" are invariant with respect to problems, they must also be invariant with respect to time. That is, they cannot depend on what era of science we happen to be in; what we know that we didn't know yesterday; what's currently cutting edge, so on and so forth. Such time-invariant phenomena are why we can always learn so much from history and especially the history of science.

This means that if we simply found the abstraction to our best methods of thinking, they would remain valid way beyond our lifetime. Problems could continue to be solved by them, the way we would have solved them, hadn't we long been gone!

So I couldn't help but ponder: why didn't Einstein choose to spend the last 20 years of his life (past all his major contributions to physics) to think about his own thinking and write down some principles or perhaps pseudo-codes to leave behind? such that they could be used to solve problems the way he would have approached them, problems that he wanted to see resolved for himself or the world.

Well, he definitely could have. But he was Einstein; he had the right to assume he could soon finish up all the things he had hoped to get done. What about the rest of us who are a lot slower than him but want to see the same progress and more nevertheless? why don't we invest a bit in subject-less thinking (and similar topics that we get to later on)?

For me, there wasn't much need for hesitation. Once again what I wanted to do for myself and the world had become the same exact thing. These observations had brought me back to the missing manual problem. Except now with the added benefit that any major progress could also be helpful to physics, eternally!

This concludes my own journey and how I got into AI and writing on intelligence, but how about all those who have come before us? Of course many have realized that the missing manual is actually a problem worth working on, just maybe not as explicitly as I have stated here. In upcoming chapters, we review a bit of what our science and philosophy disciplines have done about it and connect it to foundational issues we set out to raise.

Let's summarize what we've discussed so far:

- We tried to "rethink thinking", to put aside all our folk understandings and current meanings of the word thinking in linguistics and to start from scratch.

- A massively unconventional but simplifying approach could be to equate "thinking" to all human responsibilities whatsoever. But that doesn't constitute a definition because to define "all human responsibilities" we would end up having to define "thinking".

- We discussed that any naive attempt to define the boundaries of thinking and propose measuring the qualities of thinking would fail. Nonetheless, we know that these qualities vary, and relatively "higher" qualities must exist and do exist.

- Be it so, we described the ultimate description of our brain/mind (our thinking machine), its states, how to use it best, and so forth as the contents of a hypothetical "missing manual". This is the missing manual of thinking and cognition.

- One can make arguments to suggest that this manual must contain some abstract facts or principles for certain kinds of "optimal thinking" that are independent of human beings and various contexts, and therefore containing some kind of universality — which we called "subject-less" thinking.

- Given that we use our brain to do almost anything we do, albeit mostly autonomously, we'd of course like to "find" or gain access to an approximation to this missing manual. That is what we call the missing manual problem.

- This metaphor frames the problem for us and states the depth of answers we seek, such that we can now turn to our science and philosophy for what they may have in store for us.

Chapter 3

Artificial Manuals — Science, philosophy, and the brain

To think about foundations for intelligence, it's natural to want to step back and review the birth of AI and cognitive science. That's what we'll do in this chapter, with some focus on the psychology behind doing science including psychology itself. To help us see the bigger picture, we have to frame them as the development of some set of artificial manuals — a metaphor introduced in Appendix A: generalization of the missing manual problem.

Here's a summary of that generalization which we'll use here: If a missing manual contains the ultimate philosophies, truth, useful facts, and recipes regarding a subject matter under consideration, then the closest thing we could put together, to that missing manual, can be thought of as an artificial manual. Think of the natural one as what "God would put together" and the artificial one as our collective knowledge at any given time.

This gives us a way to talk about science and philosophy without specifying which science or which philosophy in particular! And we need that, simply because not only many different disciplines contributed to the birth of AI, but also many may have to recombine later to give rise to a proper science of intelligence. Except we don't know which ones yet and how.

This chapter may be the hardest part of the book to read, simply because it is highly multi-disciplinary. We mingle philosophy with psychology, philosophy, physics, mathematics, and computer science all in one go!

Let's start by asking ourselves a fictional question: out of all the missing manuals[35] out there, which one poses the biggest problem for us by its absence? In other words, if you could ask to access only one of the manuals which one would be your pick? Would you pick the one for chemistry? That one contains the ultimate knowledge of all chemical processes, and could help you make things, with which you could dominate the world very quickly. Or would you pick the one for financial markets? With it, you could swing any market to your favor unimaginably fast and soon hit the top of the richest-alive list.

I don't know about you but I'd ask to see the one for the mind. Simply because it would be the first one of all that we as humans would use in order to get to anything else, even to read and comprehend other manuals. Plus, with the mind-manual, you could put together an artificial manual for any other one that is missing. Therefore, missing-ness of the mind-manual is in a sense a more fundamental problem than the missing-ness of all the other manuals combined. Precisely why we chose to talk about it first.

It's more than fair to still be on the fence on identifying it as a problem. No one ever says "oh, my life is so crippled by the missing manual problem". The reason is that we are actually all intuitively aware of the issue. We just deal with it naturally enough that we never get to formalize it and spit it out in words. Let's see how we come up with some intuitive self-manual by reviewing the developmental psychology behind it. That will in turn give us hints about our collective psychology in approaching new disciplines.

Intuitive Science and Philosophy

Many philosophers have long told us that our real lives are taking place inside our minds. Some like Immanuel Kant would go as far as saying even the outside world inherits its real-ness from the inside of our minds. So Kant would perhaps deny the existence of any objective reality to a missing manual

---

[35] We invent a more appropriate word called "manuon" in Appendix A. But since we have already abused the word "manual" so much, we continue to do so in this chapter.

or manuon, but even he would agree that the mind comes up with some manual of its own. In the language we have adopted, we call that forming an artificial manual. Except that it's only an intuitive manual since we cannot explicitly record it out, we can just infer its existence by observing the utility of mental activity.

Every child (even prior to birth) has to process lots of signals from the environment to figure out what's bad, what's good, what works, and what doesn't.[36] We still don't know exactly how we acquire the concept of "things and objects" but it's widely believed that it's an innate property (similar to the concept of concept, which you use in language). Evolution has done the work, has found it a useful concept, so we are born with it. Some call it the "objectness" property[37].

Given this built-in quality, right after its recognition, infants study an object relative to other objects, including their own body. First, they are bound to learn about the existence of position and motions of things. Then within a few months, about the relative properties of these attributes, gravity, and so on. After a year they understand object permanence and the 3-dimensional structure of the world quite well. That is how infants are said to become intuitive physicists! They must gain this intuitive knowledge first because the position of their body relative to other objects is the most critical piece of information for their body to deal with the external world.

Meanwhile, on a separate track, the kid is processing a lot of signals from other minds with authority around; parents, siblings, caregivers, and so on. Within 9 months they start discriminating between their parents and others. During the toddler years (1 to 3 years old), using all these signals and feedback as well as their own feelings and experiences, they build the foundations to become intuitive psychologists. Around the age of 3, they can relate to emotional distress in others and show empathy. Just like they build a model for gravity explaining the fall of unsupported objects, they form a model for the states of the mind, explaining the behavior of living objects around.

Understanding other minds is a necessity for the kid because the power to modify his/her relevant environment rests within those other minds (other agents around). These models show up later than the physics ones because they are more sophisticated (for our minds). The associated

---

[36] Binary nature of our "old" brain's innate structures (good vs bad, fight vs flight, etc.) should be considered a deep mystery of nature on its own, although most folks consider it trivial. Binary blueprints are considered the simplest form to engineer in nature. All classical computers have been built that way. If you can control the presence of any physical property vs. its absence, you're set. How can it get simpler than that? Well, the jump to preferred computation, intelligence, goals of evolution or nature is far from trivial!

[37] Granted to us by evolution to make sense of the world easier. Objectness is why children learn to count so early or why we discovered positive integer numbers before anything else like zero or negative numbers.

signals are more complex and full of exceptions. Mostly because we are given less innate ability to model them properly. Having said that, none of these issues are fully understood yet and are all under active research within developmental and cognitive psychology.

The cognitive development of kids strongly depends on these kinds of modeling acts. Any resulting model can be thought of as some combination of three types of models:

1. Robust consistent models that explain the target behavior/phenomena completely and accurately.

2. Models that have tight domains of validity. The kid knows it doesn't work everywhere and it's not fully accurate but it's a good heuristic.

3. Leftover data/observations which are not well-modeled yet. Only a mere associative knowledge of what behavior could cause what problem or bad consequence. These are kept to be paid attention to as strict rules and brute facts for the time being.

All of these start to become part of our common sense during the later parts of the preoperational stage (2 to 7 years old). Lack of a simple and consistent way to quality-model the data in part three and ad-hoc-ness of part 2, plus leftover questions in places where no clear signal is received (places where any signal would be extremely useful), all lead to the formation of intuitive philosophy!

It may start as soon as the concrete operational stage and of course continues to all later stages in life. Whether we recognize it or not we somehow form and follow some set of internal philosophies about the very many phenomena we come across and have to deal with. That allows us to judge things, including our own limits. Judging and assigning goodness and badness to things, which we appear to do naturally, wouldn't be possible without sets of internal beliefs and philosophies. Philosophies around how to be, how to think, etc.

Therefore even as a child, we are already aware of the implicit missing manual problem. We gather data, model it intuitively and form beliefs that could be thought of as intuitive artificial manuals we go by. They become second nature to us. Although we mostly keep them open to be updated, in practice it proves very hard the older we get. The more we follow them, the more they get reinforced. We all do this naturally. However, The questionable implicit bias here is towards believing that this is the best we can do! This is what the missing manual phrase questions and it is also why we have never formalized it as a problem.

Now, our approach to develop intuitive philosophies is not limited to the intuitive manuals we build regarding our own individual lives and minds, it spills over (by default) to building our non-intuitive manuals too, regarding our collective science and philosophy. That is, we can recycle the lessons from individual intuitive philosophies to learn something about our collective philosophies at the level of social phenomena, and that is of use for us.

Our formal philosophy and science are born in the curiosity of those who aren't satisfied with the artificial manuals we end up with at any given time. That can only be due to at least an intuitive belief in some utility in stepping beyond. The belief in a missing manual for the mind means more than the belief that there are brain processes that we should study and discover. That is, not only we should study how the brain works but we should take into account that there are different ways of using the brain, and there may be optimal ways of doing so.

It's true that currently there is so much we don't know about the brain that perhaps it only makes sense to focus on the basic forms of usage that are absolutely shared among almost all human brains. But that is a choice of philosophy and it's going unquestioned. Yet it has enormous implications for how the science of the brain will unfold and evolve. More on that in the next section.

Similarly, when it comes to intelligence and AI, the working assumptions towards human intelligence aren't much more sophisticated than plain statements like "humans are intelligent". Human intelligence is considered a thing on its own. Again the working assumption is that we are so far from it that we can safely approximate it as one fixed thing— much like how a distant star appears like a structure-less bright point to the naked eye. Even though we already know that it must have a lot of structure to it up close. We also know that the structure of something and its location are intuitively independent, and we could study just the structure and then use that knowledge to figure out how to get close to it.

Similarly, it may be possible that studying the structure of variations in using the brain and human intelligence could point us towards its location in the yet unknown landscape of intelligence. Again this is a philosophy that we have not even discussed let alone taken seriously.

Our default philosophy is that we should go rather linearly forward. First be able to uncover how the infants learn, and gradually build our way up from there. What if the optimal path or more efficient way to build up the science of cognition could have a backward element to it, such as leveraging some understanding of advanced thinking to extract cognitive principles that could later be used to understand infant cognitive development? Well, again that is not part of our current working philosophy of investigation.

These are important examples of interactions between science, philosophy, and our brain. They demonstrate the significance of the topic and prove a need to discuss it more formally or generally, which is exactly what we'll do in the next section. That allows us to know what we should pay more attention to when we give the account of actual historical developments.

Interaction of Science and Philosophy with the Brain

Given that this is a very widely applicable topic, let's not single out any particular science or philosophy a priori, and instead speak of artificial manuals. Specifically, let's ask what's exactly artificial about these manuals?

We may consider it artificial because *we* build it, as opposed to nature kindly giving it to us. But that's not of any use. The real question is: why should the artificial one be any different from the true missing manual? Why should it matter who builds it, us or nature? In other words, what are the obstacles to finding the missing manuals? Why could our version of the manual not be identical to the missing one?

Well, suppose that given enough time it certainly could, and ignore that enough time, maybe impractically too long. Let's turn our attention to the path to get there. In particular, the dynamics of the evolution of philosophy and science. In what follows we're going to use the word "brain" to include all about the "mind" too. This is not the convention. Many consider the brain to be part of the mind. But that would not serve us in any way, and we won't make any distinction between the mind and the brain.

Let's also put philosophy aside for a minute. Regarding science, suppose 1) we are not concerned with anything explicitly about the brain, and 2) we are only focusing on phenomena that everyone agrees to be well within the reach of human intellect. In this setting, does our brain have anything to do with the evolution of science? One may say no, not really because science mainly progresses by the kinds of experiments we are able to perform. If we have the technology for it, we will do it. We will set up experiments with no regard to the state of the theory, and by how much it may be behind or ahead of experimental observations. Of course, there are lots of interactions between theory and experiment influencing the progress of one another, but the correlation of experiment with technology is much higher than with theory. So NO, the influence of our brains on the evolution of science is better captured and subsumed by technological availability. That's one view. On the other hand, both theory and experiment are enablers for the development of new technology. And the virtuous cycle continues.

What if we are answering NO to this question because we have always given the "no" answer to the same question since the beginning of science? I believe the true answer to the question is yes, namely, our brain does influence the evolution of science directly. I am not referring to anything related to the nature of funding, availability of scientists, etc. I am referring to nothing but the intrinsic human biases towards the fundamental reality that all human brains share. This is much more than what's typically known as cognitive biases. In particular, it is about the human shape of human knowledge. You may also think of this as some ultimate cognitive bias; the bias in the very nature of being a brain and using a brain, regardless of what for. I call this bias, the fundamental bias of the brain!

That is one major obstacle that renders the artificial manuals artificial, regardless of the methods we use to achieve the manual. It's not hard to argue that we could leverage any understanding here to make the evolution of science more efficient. In particular, it allows one to augment the philosophy of philosophy and science. Any such augmentation requires one to further study the fact that all human knowledge has the human shape, regardless of whether it can be put to human words or not, but especially so.

Many philosophers have worked on this within the discipline of epistemology for centuries. Most influential are perhaps Baruch Spinoza and Immanuel Kant who dived into the subject with enormous depth. For instance, as we alluded to earlier, knowledge and objective reality under Kant are only mind-dependent. That is the Kantian mind-dependent knowledge: *mind makes up nature*!

The strong dependence of knowledge on the mind is nothing new. We know that at least ancient Greek philosophy knew about it, in what is known as anthropomorphism, where one assigns human realities to other objects, for instance, animals. The Greeks observed that the Gods conceived by people of different regions had attributes similar to the people of those regions. Visual attributes like hair color, eye color, the shape of lips, etc.

Their old conclusion that still survives, although much more refined in today's philosophy, is that any definition, expectation, or conception of reality does include some signature of its origin! In the case of human knowledge, it would be the human realities. That is to say that all human knowledge has a human shape!

As we said, philosophy has considered its own interaction with the mind. What about science and its evolution? That is largely (if not completely) ignored and that's where scientists have perhaps fallen short. The most sound and consequential example by far, in all of the history of science, is manifest in nowhere but the evolution of the foundations of quantum mechanics.

If you are not familiar with quantum mechanics, no problem. All you need to know is that there is a conventional mainstream theory of quantum mechanics that people use in order to calculate, predict, and explain things, and it works flawlessly with jaw-dropping accuracy. Now if you want to really understand it, which is to know why it is that way and not any other way, you get into trouble. The conventional theory that almost everyone is taught says that's the wrong thing to want to know. It is an axiom, in particular, the very first axiom of conventional quantum mechanics: Your knowledge is your measurement. Period. And if you follow that, you won't have a problem doing physics.[38]

OK. Where is the interaction with the brain? If you guessed that conventional quantum mechanics didn't become conventional overnight, you're right. There were lots of debates that are still not fully settled. The very fact that they aren't fully settled has a lot to do with what we are talking about in this very section. Instead of going after various interpretations of quantum mechanics, which would be the most popular way of discussing the topics, we are going to take a different route. Let us briefly discuss how tricky dealing with our fundamental biases is. In this case, with regards to the evolution of our understanding of quantum mechanics.

Niles Bohr is considered the father of standard quantum mechanics mainly because he was the central force that made conventional quantum mechanics stick. He was successful in doing that because he was successful at silencing the concerns of the giants regarding the nature of quantum theory. Namely, those of Max Planck, Albert Einstein, and Erwin Schrodinger, who initiated the birth of quantum mechanics. How did Bohr do that? Well, simply put, he used the existence of human biases against them, saying their arguments felt valid only because it's just humans preferring

---

[38] Well, until you get into gravity, space-time and working on serious reconciliations there, attempting at either gravitizing quantum mechanics or quantizing gravity (quantum gravity).

human realities! Very clever debate-cutter argument, right? Cause no human can stop being a human.

Without going into any technical details here, we now understand that both sides of the argument were somewhat misunderstood, and also both sides of the argument had flaws due to this interaction of the brain with science. What do I mean by misunderstood?

On Einstein's side, people heavily misinterpreted him. By misusing his statements such as "God doesn't play dice" against him. As if, Einstein had any trouble understanding probabilistic theories. For Einstein, "locality", a solid concept well defined by his theory of relativity, was what was at stake. In retrospect, the only thing that can be potentially considered a flaw in Einstein's argument was this insisting too much on "locality" being a fundamental element of reality. My most favorite paper of all time is the ingenious paper co-authored by Einstein, named EPR[39], that contains the famous phrase: "no reasonable definition of reality can be expected to permit this", where I believe "this" refers to violation of Einstein's locality.

On Bohr's side, contrary to what most thought, his exact philosophical standpoint wasn't that reality should depend on measurements, but that EPR's expectation regarding what is a reasonable definition of reality, is only based on *human* realities! This is not an unreasonable position by Bohr, and that's why it worked on silencing people. After all, human beings are insistent on human realities. However, the existence of this bias shouldn't mean that it can't point us anywhere useful. As long as we're aware of the bias, we can follow it and see where it takes us. In the case of EPR, following it would have been to construct a theory that would be called "locally realistic", that is a theory of local hidden variables. The title of the EPR paper is "Can quantum-mechanical description of physical reality be considered complete?" and in it, they tried to argue that no, namely, quantum theory is not complete, with the hope that something could be added, perhaps some hidden variables, that could explain away the peculiarities of quantum theory. No one tried seriously to do that for almost 30 years until John Bell, who published a testable consequence of a "hidden local reality" in 1964. The so-called Bell inequality. But even Bell didn't work right off the EPR paper. Rather, he got his inspiration and conviction from the work of David Bohm who had constructed a different version of quantum mechanics, a deterministic but non-local theory.[40]

---

[39] Paper by Einstein, Podolsky, and Rosen, known as EPR.

[40] Bohm himself had been encouraged directly by Einstein to pursue a different formulation of quantum mechanics and he did just that, except he ended up with something opposite to the kind of formulation

The point is that Niles Bohr and folks in his camp, instead of dismissing EPR philosophically, could have constructed what it suggested, and derived a testable consequence just like John Bell did. Regardless of whether they believed quantum mechanics would violate the consequence or not. Let's also keep in mind that no necessary ingredient to do that was developed during those 30 years prior to John Bell's work. Similarly, Einstein himself was more than capable of constructing a hidden variable theory and finding bounds on statistics of measurements on systems that he himself proposed in EPR, i.e. "entangled" pairs of particles. But he didn't! Still, what EPR ignited is now the foundation of many quantum-based technological projects such as quantum communication and cryptography.

This observation just shows that both sides were affected by the fundamental biases of our brain. Einstein, rightfully so, by being married to locality and space-time which are of course reality to a human brain since all our brain processes are made of these things. And Bohr, by being too eager to eliminate anthropomorphic bias prematurely.[41] Our takeaway is that knowing about the bias alone isn't enough, dealing with situations in science where these biases play a role is very tricky and their interaction with the evolution of science is far from simple and therefore deserves much more contemplation than so far received.

The example of quantum mechanics is arguably the best illustration of anthropomorphic biases interacting with the evolution of science. Having said that, because it's still too recent of an example and not everything and everyone is squared with quantum mechanics, especially when things get to gravity, I will give you another example to make affairs a bit more obvious. This time an old and well-settled case. That is, why did it take so long to go from Newtonian mechanics to Einstein's relativity? Short answer: Newton's God!

School of sciences opens up with Isaac Newton and his Principia. Not that no one did science before, but he brought around the grand revolution, going from natural philosophy to physics. To this day, almost any science curriculum includes Newton's "laws of motion" quite early on, it's physics 101! It forms the perception of what quality science is for many kids. Newton's formulation says that basically you can accelerate an object to increase its speed indefinitely. So, in principle, the

---

Einstein has hoped for. Bohm's formulation is explicitly non-local and has origins in De Broglie's pilot-wave theory.

[41] I tend to think that Bohr's way of biting the bullet was too philosophically twisted and maybe intentionally so to take it somewhere that further argument with him would be too challenging. Though I don't think anyone knows that for sure one way or another.

speed of the object can reach infinity. But Newton being both a mathematician and a physicist, knew that physical infinity is just a large number, a much larger number than anything else you care about in your particular system under analysis, as opposed to the abstract notion of literal infinity.

So potentially he could have said, well, infinite speed is just absurd and my formulation should only work when speeds are not too large. But had he said that (which he should have at least in his own mind), he would have almost certainly proceeded to consider a very large but finite speed limit. Whatever the limit. Just too large compared to perhaps celestial speeds measurable at his time. At his time, there was already a theory of relativity and that is Galilean relativity. The theory of Galileo Galilei, the Italian astronomer, physicist, and the father of the scientific method. You don't need to know Galileo's theory, just that it was there and well-tested 30 years before Newton's work. All Newton had to do was: grab Galilean relativity and put a limit on maximum speed, cause let's face it, it cannot be literally infinite. To stop speed from ever becoming infinite for any observer (in any reference frame), he would have no reasonable choice but to assume the same speed limit for all observers. Otherwise, if different observers had different speed limits, they could build upon one another and it would be eventually possible to reach infinite speed, which again would not make sense. So just this one assumption alone would let Newton turn Galilean transformation to Lorentz transformation, the core of Einstein's theory of Special Relativity. These are transformations that take you from the observations of one reference frame to that of another reference frame. They tell you how the space and time of one observer are related to the space and time of another. All the obscure effects due to Special Relativity you hear about, like length contraction, time dilation, etc. come from these transformations that tell you how measurements of different observers are to be translated to one another.

What I am trying to highlight is that Isaac Newton should have been able to come up with some version of Einstein's relativity very soon after his own theory of motion. No need to know about electromagnetism, the nature of light, so on and so forth, none of the things that were developed over the next two centuries after him. Just like John Bell didn't need to use anything that was developed during the 30 years from 1935 (EPR) to 1965 (Bell's inequality) for his work. Before we jump into why Newton didn't do what he easily could have done, let me mention how relativity eventually came about.

It was the infamous Michelson-Morley during the 1880s which showed that the speed of light in different directions was the same. And the open question was how could the aether, the medium that was previously postulated to be there just to allow light to travel, could exist, if its speed's direction could not affect light's speed. That was not consistent with the existing theory of relativity, namely that of Galileo. None of this was resolved until Einstein came about and resolved it by his theory of special relativity in 1905, which says the speed of light is THE speed limit. But again, in retrospect, Newton could have just put any limit on speed (not necessarily light's speed) and still derive interesting relativistic effects and attach them to his Principia.

The fact that there is not even a discussion along those lines in Newton's Principia only means he didn't quite question it in his mind. Although he was troubled a bit by instantaneous action at a distance in his theory of gravity, which could give hints on trouble with infinite speed, those don't count as contemplations on relativity. Newton was quite a religious man, he believed in an "infinite and eternal" divine power, God, the source of infinity. But he didn't think of God as any kind of "matter", he was explicitly associating the infinity with the existence of infinite space extending in all directions. In his imagination space didn't need matter for it to exist, just God. Matter was a finite play for God, confined to a finite region of the otherwise infinite space. So his jump to infinite speed of matter is not justified even logically using his God as an axiom. Newton must have been unaware of his bias towards infinity because of his belief in God. This seems to be the only explanation as to why he jumped over questioning the absurdity of infinite speed against his scientific reasoning but not that of gravity's instant action at a distance.[42]

Furthermore, creation, causality (which we will cover in-depth later), respecting higher power, and what you can't do yourself, are all only human concepts and realities. The tough challenge of making the distinction between these and some human-independent reality is due to the same fundamentally anthropomorphic bias we have been considering in this section. In the case of Newton, it resulted in the conception of a God mirroring his beliefs, just a more sophisticated version of the anthropomorphically-biased version of ancient Gods. It might be tempting to think of Newton's God as a more scientifically-founded God relative to those imagined by the ancients. But even then, one can easily argue that invoking God in science could be much more useful to explain a mysteriously special and finite non-zero value, than infinity or zero. These are neither theological statements nor any statement about scientific boundaries. We are simply pointing out that infinity or zero in basic science are much easier to find satisfying explanations for, than a finite non-zero value for some "irreducible" fundamental variable.

---

[42] His concerns regarding gravity were addressed (with Field Theory which makes all interactions manifestly local) way before Einstein's relativity came about.

In Einstein's general relativity because space and gravitational fields are not separate things, you can get rid of this issue of infinity of space or Newton's God's sensorium, by appealing to what curvature can do. That is the curvature of space. Think of a circle and suppose you're strictly bound to live on it. You can keep lapsing around it forever and it would not be a bothersome infinity because at any finite time you've gone finite rounds around it.

It is worth noting however that the problem with infinity of space is still with us in other regards. But this time it is manifest and tied with indirect observations. For the first time in the history of physics, people are contemplating a physical literal infinity. That is introduced by what is known as "eternal inflation" based on the theory of inflation. Inflationary cosmologies are currently the most popular theories in early universe cosmology. Theories with explicit exponential functions responsible for the fast expansion of space in the early universe. The core theory was Introduced around the early 80s and up to now has been able to adapt itself to be not incompatible with experimental observations. Once the inflationary process kicks in, it's hard too hard to stop it, as portrayed by eternal inflation. Thereby, not only it would introduce and require literally infinite space but also infinitely many universes within this infinite space — a multiverse.

So that's one reason that encourages people to believe in a multiverse. Once you believe in this kind of multiverse, no other multiverse should sound too crazy to you. That's why many inflationary cosmologists wouldn't call the infinity of space a problem but rather, a solution. Can you put a limit on that infinity and get anything right? Unfortunately, it's not nearly as easy as the fix for the infinite speed of Newton's theory. It requires a much more radical move. Many brilliant people are actively working on these problems and there are already many hints that the satisfactory solution would likely come with abandoning the notion of space (of all possible positions) as a fundamental concept. Such progress would force people to also abandon a solely reductionist mentality which has fundamental roots in spatial thinking (in a Euclidean sense which we are all intuitively familiar with).

You may wonder that even if space isn't fundamental, how could this help remove the infinity? Well, when something isn't fundamental, it has a bounded domain of validity and applicability by definition. That finite domain would be determined by the more fundamental concepts whatever they may be. Here's an analogy to one possible way it could. Imagine an infinite space with one finite closed "box" in it. Further, imagine that the only light sources in this space are located inside the box. One can think of the shadow of the box to be literally infinite. But that's a problem with the language of shadow being applied where it shouldn't be. That is, an observer of any shadow should be able to receive some light from the very same light source causing the shadow, or else the observer is entirely part of the shadow and can't perform any experiment to distinguish itself from the shadow. So therefore this infinity isn't a problem. One would just have to talk about the light instead of the shadow. I am not at all suggesting that this maps to how the problem with infinity of space will eventually go away. But just that in a similar manner, it could go away when we switch from talking about the space of positions and how it expands, to what manifests space!

Our understanding is bound to increase. Not just because historically that's been the case but because of the current state of affairs and interesting puzzles that are right in front of us. The point here is that we may be able to do that more efficiently if we adopt and augment our philosophy of science to what's in demand now. In particular to take the human shape of human knowledge more seriously. If Newton and Einstein aren't immune to it, none of us are. However, saying that these are all human biases and we should remove them isn't the full answer either. Because we can't quite just remove them. As we saw earlier, that's what Niles Bohr tried to do. Jumping to remove such biases by totally ignoring them. This would introduce its own bias, a worse bias. Nothing slows down or stops the progress of science like adopting the loose "anything goes" mentality, saying whatever we insist on is perhaps because of human biases. That's a real show-stopper. A dangerous thought that unfortunately is getting prevalent in parts of our society. The ill-informed thought that because science doesn't know everything, perhaps it's OK to doubt everything.

If we can't remove the bias, what should we do about it? Well, the best first step would be to become aware of it. And that starts by asking the right questions to expose it. What kind of questions? Historical examples we reviewed here should give us good hints about what we should watch out for. I hope that by now you find it the case that, if we are actually concerned with the evolution of science or have any hope in a scientific ability in predicting the evolution of science, we have to come up with a slightly different theory. A theory in which philosophy, science, and our brain are not separate non-interacting elements.

We have to notice that so far we have ignored the correlations and causation between these elements and that's only an approximation in a theory of evolution of science. An approximation that is too crude to explain observations. We are not even close to such a theory and the claim is that constructing such a theory is not possible without a new philosophy of science.

Going forward, the correlations are likely going to be even stronger, and separating the fundamentals of the brain from science and society will in hindsight be viewed as a giant historical ignorance. Of course we know that we don't understand the brain and how it works. The peculiar thought here is that we don't understand that we don't understand the brain (an unknown-unknown). Meaning, we are likely already experiencing the consequences of such unknowns without being able to pinpoint and attribute them to our ignorance about the brain.

Strong correlations or not, to go towards a theory of evolution of science and find opportunities to increase efficiency, one must consider a philosophy of science in which the interaction of science with the brain and its philosophies are accounted for. This observation suggests that what could help with gaining awareness of our collective biases is

- questioning psychological motivations behind scientific moves and assumptions
- questioning the nature of produced knowledge, and
- questioning the initial philosophical viewpoint and comparing it to the resulting philosophies, which are typically developed long after paradigm-shifting scientific results are reported.

That is exactly what we'll try to do next in our very brief review of the history behind AI and cognitive science. That's much more important than listing all the events, programs, and apparent successes or failures.

Lastly, before we leave this section, let's note where the fundamental bias is playing its most acute role. That is in the philosophy of mind and philosophy of cognitive science, where the power of the bias reaches its maximum potency. That's its home base. When the mind wants to talk about itself. Those who have been taking their sole introspections too seriously are perhaps the most vulnerable to find their judgment farthest from the realities we stand to eventually uncover.

A Brief History of AI's Parents

Perhaps no field other than psychology has contributed more to the early foundations of intelligence. Computer scientists do have some practical definitions for intelligent agents, and we'll review those later. But when it comes to the nature of intelligence, core ideas have mainly been shaped in psychology and to some extent with the philosophies of mind. So it's appropriate to start our discussion by reviewing what "psychology" is and isn't before we get to other fields like computer science, neuroscience, cognitive science, physics, and mathematics, which have all influenced AI's foundations.

By literal translation, psychology is supposed to be "the study of soul". That's how it was defined in ancient times, perhaps when everyone believed in some form of soul transcending the body yet influencing it (and its behavior) in magical, mysterious ways. It was not until the late 1800s, that psychology was established as a scientific discipline. At that time physics was already an icon for rigorous science; a discipline with precisely defined and strict rules giving rise to powerful predictions and unforgiving accuracy. It was only natural that some people wanted (and still do) to study the mind in a similar fashion. This is an instance of what scholars of science outside of physics call, "*physics envy*". Envying the mathematical precision, simplicity, comprehensiveness, and precise interaction between all the ideas and theories in physics that is lacking in other disciplines including biology, social sciences, economics, psychology, and now in AI.

The desire to understand the mind, the obvious importance to make predictions about it, along with physics-envy shaped the first generation of work in psychology. It shouldn't be a surprise that the majority of early work in psychology tried to imitate Newton's practice. That is to follow Newtonianism, a natural philosophy with the hallmark that the world has a set of rational and comprehensible laws, we should look around, gather observations, and write down the underlying "laws". That's what early psychologists tried to do, literally imitating Newton, except in a different context.

Prior to that, theories of mental phenomena were just too absurd. Take Phrenology for instance. If you haven't heard of it, that's because it's a dead pseudoscience. At the end of the 18th century and the beginning of the 19th century, some physicians and anatomists believed there exists some set of perfect skulls; shape, size, weights of heads. The practice of phrenology involved measuring the contour of a skull, paying particular attention to any bumps deviating from "perfect head shapes" to predict mental traits, characters, and potential psychological disorders. If this sounds too absurd, maybe it is so only in hindsight. They did at least narrowed down what most influences character, thoughts, and emotions to specific areas of the brain. One should contrast that with Aristotle's conclusions that consciousness resides in the heart!

Later in the 19th century, most people considered these phrenological practices as totally deviating from science. Fast forward towards the end of the 19th century, some people were coming up with theories laying out 100s of laws to describe various cognitive and behavioral phenomena, and exceptions to those, etc. To explain and predict all the same things that phrenology wanted to predict, and many things that psychology still does. Back then what came to be in fashion was to come up with some set of laws, like a numbered list of laws similar to those of Newton's laws of motion.

Following Newton's well-known example, people wanted to speak only in terms of quantifiable and potentially measurable variables. So they focused on observable inputs and outputs to the mind and deemed all other theoretical treatments as non-scientific. This quickly led to the first dominant wave in psychology: *behaviorism*. Treat the mind as a system, and what you can observe of it is only the external behavior it results in. Focus on finding the laws that predict these observables, i.e. the behavior, and you're done. John Watson was the one who officially established the psychological school of behaviorism, drawing inspiration from the work of the infamous Ivan Pavlov. The two main iconic figures of behaviorism were Pavlov and B.F. Skinner. Skinner, who was named the most influential psychologist of the 20th century, came a bit later and was inspired by both Pavlov and Watson.

Pavlov came up with what's known as *Classical Conditioning* and Skinner with *Operant Conditioning*. Both of which are theories of learning in psychology, in particular, they are forms of associative learning. How does psychology define learning? Simply put, learning is a process of acquiring enduring and behavior-affecting information, through experience. Associative learning, more specifically, requires a direct link between certain events or external stimuli, and certain behavior. These links get formed in a process called conditioning, and there are various forms, underlying theories, and definitions for this process.

Because the experiments (on conditioning) involved repeated sets of trials of presenting a stimulus (or removing one) to the same subject until the subject is "conditioned", the dominant theories explaining these observations involved concepts of (positive or negative) reinforcements, (primary vs conditioned) reinforcer, reinforcer schedules, and so on.

Fast forward many years, they are the basis of perhaps the most attention-grabbing AI algorithms, namely, Reinforcement-Learning algorithms. We'll talk about that in the next chapter but this learning is not to be confused with the psychological notion. Most AI researchers consider the work of Pavlov, Skinner, and others alike in that era, as obsolete and not worthy of contemplation. Some tend to now portray Pavlov as the guy who shakes a bell to make the dog drool or Skinner as the guy who puts children in cages for his experiments.

Yet, there are psychologists who have remained curious about associative learning and find it possible that we don't fully understand these old experiments, and that the theories and conclusions we walked away with are not totally correct. For instance, re-examining the conditions in which these experiments took place, and putting them together with our cutting edge understanding of how memory works, some conclude that perhaps the dog was conditioning Pavlov and not the other way around! Arguing based on the fact that many of these experiments took place in the same environment, and one should take these lab environments into account as crucial confounding (think: causal) variables. And why would the dog forget anything at all? And why is there a need for so many repetitions? There are arguments in favor of different kinds of answers to these questions and we won't be settling anything here, but perhaps associative learning could still reveal important lessons for AI provided that we are open to question some of our early assumptions.

Some of the greatest examples of Newton-imitating psychological theories took place during this era of behaviorism. In the early 1940s Clark Hull published two books on learning and behavior. His work included more than 100 laws and postulates laid out in an axiomatic system. His book is full of mathematical equations quantifying various stimuli and responses. Here's an expert from his conclusions in his 1943 "Principles of Behaviour":

"... If these three tendencies [referring to tendencies to see behavioral sciences as natural sciences, to replace theological and folk talk about them with mathematical statements at every possible point] continue to increase, as seems likely, there is good reason to hope that the behavioral sciences will presently display a development comparable to that manifested by the physical sciences in the age of Copernicus, Kepler, Galileo, and Newton."

Hull was among the early thinkers who believed we can replicate mental processes in some machines.[43] It would not be far-fetched to assume that his attempt was to write a psychology version of Newton's Principia Mathematica. Needless to say, despite his admirable pursuit, his work was not a success, at least as measured by whether or not people followed up on it. His critics considered his work impossible to use.

Skinner's work enjoyed all the attention instead. Was that fair? Or better posed, was it true that nothing was right with Hull's work and all right with Skinner's work? Was it right to assume that psychology is better off without holistic mathematical theories? No one can give a defendable solid yes or no answer to this question. Why? The reason is that the answer is outside of science, namely because we lack the right philosophy of science for that in the sense we discussed in the last section.

Marvin Minsky had some answers to this question. He used to argue that laws and purely mathematical theories work for physics cause it's simple there, psychology isn't simple at all and it

---

[43] Encyclopedia of Psychology, 2000, Oxford University Press.

doesn't admit such theories. Basically, his recipe was: just don't look for that simplicity and mathematical laws when it comes to psychology and cognitive science. That's more or less the overall attitude of almost everyone in the field anyway. Newtonianism did give rise to the modern philosophy of science but as you see, it leaves many issues unaddressed and many questions unanswered. We'll pick that up more in the next section.

Let's get back to our review of developments in Psychology. Of course, behaviorism was not the only game in town, multiple schools of thought were present at the same time and still is. Psychology has always been a very diverse and dynamic field. That's with respect to both the types of questions that are asked within the field and also the methods with which to go about answering them. Today's definition of Psychology is as follows: the science of behavior and mental processes. How did mental processes come into the definition here? Didn't behaviorism deem it "unscientific" already?

The walls of behaviorism started to crack with the emphasis of Sigmund Freud's work on subconscious processes and the subsequent establishment of his school of psycho-analysis. Still, the behaviorist's flag outlived many others. Carried on especially by B.F. Skinner, who in 1957 published "Verbal Behavior". In which, he doubled down on the partial successes of his operant conditioning and used it to claim that it also explains language acquisition in children. Two years later, Noam Chomsky published a critical review of Skinner's work arguing strongly that behaviorism and mere environmental and experiential learning cannot explain language acquisition and that children are born with some innate structures regarding language and grammar.

Modified residues of these arguments are still around within AI and cognitive science as to whether it's best to think of the mind of human babies as blank slates to learn from experience or as one with many innate structures to facilitate learning. Of course, there is a spectrum of ideas around and we'll come back to this in the next chapter. For now, the point is that these movements, most notably by Chomsky, were among the last hits to bring down the walls of behaviorism. This led to what's known as the cognitive revolution towards the end of the 1950s.

There were other crucial developments too behind the cognitive revolution. But before we get to those and what the revolution was about, let's mention that when it comes to intelligence, psychology doesn't say much except in the context of Developmental Psychology. Even then all the statements are based on very relative or subjective definitions of intelligence. Relative to some mean or median of a sample population, relative to what's considered normal or majority's position. Everyone's either heard of or taken an IQ test. IQ scores are in fact aggregates of a series of standardized tests, most common among those are Wechsler Adult Intelligence Scale (WAIS) and the Wechsler Intelligence Scale for Children (WISC). The current version of WAIS for instance has 15 different subtests. There are many other tests too and yes, psychologists are aware of the existence of an overwhelming variety of biases here. From subjectiveness of some measures to how to administer a test in the presence of stereotype threats and so on. Developing new tests for measuring intelligence is still part of psychologists' agenda but the main framework stays the same. There are no absolute measures, it will be a test taken by many individuals and some arguments will be made based on the resulting statistical distribution. The non-controversial conclusion is that intelligence testing in humans remains a giant mess!

In the absence of a satisfying science telling us what intelligence is and isn't and quantifying it, what did most people think about the ultimate signs of intelligence? That is, most people in the pre-1950s. To answer this question, we should think about how things were back in the pre-computer era. Or in the era when "computer" used to be the title for a person.

This is where we are going to use our conclusions from the previous section. Let's question the beliefs of folks of a century ago. What was the dominant mindset? Well, homo sapiens believe that they are more intelligent than other species and animals, evidence being that they have a much more sophisticated language, they make tools and sophisticated plans, etc. So they correlated intelligence with the difficulty of an intellectual task. What were the hard intellectual tasks for people? The hardest thing for people was (and perhaps still is) doing abstract logic and math. The most intelligent people they knew happened to be in this class, such as mathematicians, logicians, philosophers, and alike. Everyone believed that you got to be highly intelligent to do these things. In fact, if you were able to do that, no one could say that you're not intelligent. Not just because it's hard for most people but also because it was believed that you needed logical reasoning to figure out anything useful towards accomplishing non-trivial goals.

No one gives you intelligence-credit if you randomly stumble upon something that works, and to prove that you know what you're doing you must present your logic and that's formalized in math. Moreover, as we discuss below, computing machines came out of our work in abstract math and logic, so the potential intelligence of machines would be illustrated best if they could do just that. Doing what their creators were doing, i.e., abstract logic! This mindset turns out to write so much of the history of AI in the last century, and not in the most positive way. Some of the current debates in AI are reminiscent of this old bias too, coming directly from the interaction of our brain with science.

Let's turn our attention now to the state of mathematics at the beginning of the last century. The most influential mathematician at the time was David Hilbert, who had launched a program at the time to formalize all of mathematics.[44] Formalized as in formal languages with some predefined alphabet, a fixed set of axioms, and rules to come up with new valid statements. Prior to the 20th century, foundational areas of mathematics were fragmented. There wasn't a cohesive foundation of mathematics. You had various mathematical truths and theories based on different sets of axioms. There were areas of mathematics that were not even fully "formalized". There were attempts to remedy this situation but they started to run into some paradoxes.[45] The so-called Hilbert's program was proposed to overcome these challenges.

It is fair to now characterize Hilbert's desire to come up with a "grand unified theory" of mathematics if you will. To him, mathematics was "freedom from contradiction" and that includes a lack of contradiction between different areas of math. Therefore, it should be possible to put everything under one roof. That is to axiomatize everything, put them together, and find the minimal set of axioms such that, not only it's all consistent with itself but also it would be consistent with any higher-level mathematical system that could come up later (since it would have to emerge from this hypothetical foundation). That was Hilbert's program in a simplified nutshell. He was already on that route back in 1899, when he published "foundations of geometry". A work on the axiomatization of geometry. He came up with his own set of axioms replacing Euclid's (axioms of Euclidean geometry) and kept revising it for the following decade or so.

Here comes the big assumption by Hilbert. Until 1930, he did not distinguish between the validity of a mathematical statement and whether or not one can prove it. To him, there was no such a thing as an unsolvable problem. If it's valid, it's derivable (from the set of axioms that Hilbert's program was after). Now that we have computer science as a discipline, we speak of computability and not derivability. But how did we get here?

In 1931, Kurt Gödel in his Ph.D. thesis effectively shattered any hope in the success of Hilbert's program. He published an incompleteness theorem that shows any self-consistent formal system carrying some finite set of axioms admits some statements that are valid but not provable based on that set of axioms. You can be free of inconsistency only if you give up on provability. Basically:

---

[44] Known as Hilbert's program, whose goal was to capture all mathematical theories by a finite, complete set of axioms, along with a proof that the axioms were consistent. That would give an irrefutable foundation to all of mathematics. This program was proved hopeless by Gödel in 1931.

[45] See for instance Russell's paradox in set theory.

forget about finding an axiomatic system for all of mathematics, not going to happen. This was the beginning of such a significant development that still hasn't been fully absorbed by science as a whole! More on that in Volume II.

Gödel broke the equivalence between derivability and validity. The novelty of this perspective shook everyone. It meant that not everything true could be derived or computed. His work was a massive influence on Alan Turning and Alonzo Church, who both independently wanted to understand what are these computable or derivable things in mathematics and what's not included in it. So each independently came up with their own formulation of computables. Turing's work proved to be more influential because he framed the question slightly better!

If a statement is derivable, then there must exist an effective procedure to derive it. And once you have a procedure, then that procedure can effectively define a machine, namely one that follows the procedure. Following a procedure was exactly what the people who used to be called computers did. So Turing laser-focused on giving a formal definition to this procedure of computing. He observed that it's basically nothing but starting from a sequence of symbols, acting on them based on some rules, turning them into another sequence of symbols, repeating until a halting condition is met, and returning the final sequence. The formal definition is known as the Turing machine. Turing proved that a general version of this machine can do anything that any other Turing machine can do, and for that reason, it's called a Universal Turing Machine.

But that was not sufficient to confirm compatibility with Gödel's work. Even with assuming that Turing machines can compute everything computable, Turing still had to come up with at least one statement that would be valid, yet could not be computed by his theoretical machine.

Enter the halting problem. Suppose one is operating a Turing machine. Perhaps the most useful for an operator of any machine would be to know how long it's going to take before a certain procedure halts (even our brain is now known to first consider the duration of any task or goal, during any awake mental activity about that task or goal). Even more basic than that whether the procedure is ever going to come to a halt. Turing observed that no procedure can spit out a yes or no answer to this question for an arbitrary procedure with some arbitrary input. You would have to actually run it to know whether it halts or not. This is the first problem that was found to be not solvable by a Turing Machine. It was sufficient of an answer for Turing in response to what can't be solved with his machine. It just seemed like a perfect example of statements that are valid but not computable, the ones Gödel proved their existence. In this case, a Turing machine would loop forever, i.e. fails to compute an answer.

There is a deep connection between Gödel's theorems and Turing's halting problem. They aren't the same thing but one cannot exist without the other. Gödel-type statements can be used to prove the "undecidability" of the halting problem and Turing machines (with their halting problem) can be used to prove Gödel's incompleteness theorems. Yet, there is still an assumption here and that's that everything computable could be computed by a universal Turing machine. Alonzo Church made the same assumption about his set of computables (equivalent to that of Turing).[46] Their assumption is known as the Church-Turing thesis. And much of computer science is founded on this very assumption.

On a different thread, advances in physics and electrical engineering allowed engineering computers that in principle had the capabilities of a universal Turing machine. Most of the design credit here goes back to the wizard mathematician, John Von Neumann. So now we had computers that were designed by logicians based on mathematical logic and to do logic. The influence of mathematical logic on the origination of computer science left such a strong mark on the field that almost 70 years later is still visible. If machines can do logic, what can they NOT do?

Well, a group of highly accomplished individuals said nothing. John McCarthy and a few others proposed "a 2-month, 10-man study of artificial intelligence be carried out during the summer of 1956 in Dartmouth College" to make a "significant advance" towards figuring out how to make machines "solve kinds of problems now reserved for humans". These are the words from their proposal to Rockefeller Foundation, the sponsor of the summer workshop which ended up running for six weeks. The term "artificial intelligence" was stuck from then on.[47] The participants included Claude Shannon, Marvin Minsky, Ray Solomonoff, Herbert Simon, and several other big names. The mindset was that logic is perhaps the most intelligent thing an individual human does, so from there on it shouldn't be far to get to full-fledged intelligent machines, replicating anything a human can do intellectually.

---

[46] Church and Turing use different models of computation. While Turing uses a sequential model that is easily imaginable mechanically, Church uses a functional model called lambda calculus, which can be used to simulate any Turing machine. Conversely, Turing machines can simulate the computation in any lambda calculus.

[47] In fact a lot of inspirations for replicating intelligent behavior in machines existed already in other programs at the time, most notably of which was Cybernetics of Norbert Wiener at MIT, with a big focus on control theory. The two programs, AI and Cybernetics, had some uneasy coexistence but thanks to the super-sexy name and perceptions of the term AI compared to at least "Cybernetics", AI got lots of funding to come to dominate the stage.

Just to get a sense of the atmosphere in those seminars, I just refer to a main highlight of the workshop. That is, a computer program was demonstrated to prove the correctness of a bunch of theorems in predicate logic. It was named the "General Problem Solver". Our focus here is on the psychology of the attendees and such chosen names are among the best sources for that. To almost everyone today calling anything a "general problem solver" is a laughable act. Of course, we are using the power of hindsight here a bit unfairly, but it helps us conclude that all this logical talk about logic-based intelligence in machines was nothing but anthropomorphic-like emotions taking over the better part of a few brilliant individuals.

What about us humans? How do we perform intellectual tasks? They knew that logic is never practically used in a vacuum. It is used on or applied to things we already know about the world, to some piece of knowledge. Well, just in time, advances in brain science and the field of microscopy allowed our understanding of neurons and human nervous systems to get to the next level. The next level was to become mathematical. The field of Neuroscience was established. We had understood enough about the anatomy of neurons and their electrical character to describe their function mathematically and to explain how humans may be performing intellectual tasks. Models attempting to explain the full function of neurons were developed as early as in 1952.[48] That was when action-potential was introduced and firing or not firing of a neuron was viewed as a discrete event indisputably carrying a piece of digital information.

I said just in time because quite unrelated to neuroscience, the field of processing electrical signals (by viewing them as encoding information) was being rapidly developed at the time too, thanks to Claude Shannon's mathematical framework. Shannon established the field of information theory by his 1948 paper on the subject. He not only introduces the field formally in his paper but also solves the most important problems in it already!

These developments in 1) electrical engineering and information theory, 2) neuroscience and 3) computer science, were together a mighty force that took the world of studying cognition by a heavy storm. The storm brought the language of computation, information, and computation as information processing, into the world of studying cognition. It introduced what's known as the computational theory of mind. We'll open this further when we discuss philosophies of mind. But

---

[48] Hodgkin-Huxley 1952 model introduced the action potential in a mathematical model for transmission of electrical signals in neurons. However, almost 70 years later, we still don't have a model of a neuron or neurons that describes everything about them, mapping to their full reality.

basically the belief here is that the mind is just a specific kind of computer program on a specific kind of hardware, the brain. This philosophy is called computationalism.

Combine these with the developments and shifts in attitude in psychology (away from behaviorism), you get the gist of the cognitive revolution. It's known as cognitivism, a response to behaviorism: We can and should understand the mind and cognition on its own, no need for observable behaviors. And there is a way to do that already, namely what's put forth by computationalism. There seemed to be a need to establish a whole new discipline that would shape its own identity soon. One that of course overlaps a lot with the existing disciplines of psychology, linguistics, and anthropology, but also with neuroscience, computer science, cybernetics, and AI.

What name would you give such a discipline? It wasn't a straightforward choice. Quite initially, names like information-processing psychology were being thrown around before settling on *Cognitive Science*. Cognitivism needed a core philosophy like computationalism to stand on. Indeed, computationalism was the only defendable philosophy in cognitive science all the way till the 90s and still remains the dominant one to date. We briefly mention the other more recent views (referred to as *post*-cognitivism) when we cover philosophies of mind but they don't change any of our conclusions here.

So Neuroscience, Cognitive Science, Computer Science, and AI were all born basically in the same decade! That requires a pause and ponder at the very least. One can understand this fact much better through the lens of the fundamental bias we discussed in the last section. I leave that discussion to the diligent reader. Since its advent, computer science has included the goal of doing all the things that humans can do. The biases of the founders and other pioneers led them to consider logic and reasoning as the hard problems and set the course for the kinds of dynamics which later resulted in controversial developments and forced the field to go through several "AI winters".

Still, there are debates around AI reminiscent of these biases. That being said, no one is any longer confused about the non-triviality of AI problems outside of the domain of logical reasoning. That confusion was over at least two, three decades ago. Many researchers by 1990 came to the realization that those logic-represented problems, "the hard problems", were in fact the easy ones, while the "easy/trivial problems" (not expressed in any formal logic) are the hard ones (things like visual understanding or just common sense). Nowadays the debates on AI are more on the side of how to have a method that is general enough that it doesn't matter whether it's solving an easy, hard, or some hybrid problem.

Current State of Affairs in the Science of Cognition

At least in humans, intelligence is considered part of cognition. So let's postpone exclusive talk about AI and ask what is the current state of cognitive science. And by that I mean: do we have an artificial manual corresponding to the missing manual problem posed in the last chapter? Note that it's not about the accuracy of the content, or how close we are to a copy of the missing manual. We're also not concerned with the details of the work and research agenda of current researchers. The question of the current state here is only about the big picture, the coherency of the artificial manual, and whether we have a properly integrating and binding outline for the manual. If we have that, we trivially move towards improving the artificial manual towards ever more understanding cognition, mind, and so forth.

There are currently two main opposing views here among cognitive scientists. I present those first and then an orthogonal view. The two views are roughly as follows:

1) Cognitive science has failed to emerge as the true interdisciplinary discipline that about 60 years ago promised us it will be.

2) Who cares, we never really subscribed to that "strong cognitive science"[49] program anyway. Still good and hard work is being done under the banner of cognitive science and we're good.

The first view is articulated in-depth in a recent article published in the prestigious journal Nature by Núñez, et al. "What happened to cognitive science?".[50] In it the authors argue explicitly that the endeavor of cognitive science, the effort to combine various fields into one interdisciplinary domain to describe what's really going on in the brain, and how thought and intelligence work, has failed. The paper also questions the future of the field.

This article is responded to by others who hold the second view. Here's the concluding statement by McShane et al.'s response[51] :

---

[49] Gardner, H. "The Mind's New Science: A History of the Cognitive Revolution", Basic Books, 1987. This text elaborates on a distinction between weak and strong cognitive science, where strong refers to a new discipline that doesn't just draw from the knowledge of multiple contributing disciplines, but it emerges as a new whole and affects the boundaries of the existing disciplines and how they interact with one another.

[50] Núñez, R. et al. "What happened to cognitive science?" Nature Human Behaviour 3, 782–791 (2019).

[51] Marjorie McShan, et al. A Response to Núñez et al. 's (2019) "What Happened to Cognitive Science?" TOPICS, 2019.

"We, as practitioners in the field, are not worried about the grammatical number in the naming convention (cognitive science(s)) or whether the field is labeled as multi-disciplinary, cross-disciplinary, or interdisciplinary. What we are concerned with is exploiting the rich cross-pollination of ideas across many relevant disciplines, and training the next generation to work creatively within the space of those ideas"

They argue that we should view cognitive science as what they call an "integrating science" instead of a new science emerging on its own (the goal that only some believe in or used to, and the authors clearly don't).

In order not to get stuck in the semantics of multi-, cross-, trans- or inter-disciplinarity[52], I've chosen to use only the term interdisciplinary (with a few exceptions) in this book. In fact, whenever a group of researchers from various specializations comes together, the resulting research does have a new character. So, on an individual research work basis, the term interdisciplinary is justified. An obvious point that McShane et al. paper also refers to. We know that people are conducting good and hard work given all the constraints they have to cope with.

Here, however, we are not concerned with what individual researchers are doing or to sit and judge comfortably from afar while practitioners are wrestling with very hard to form research questions in the trenches of the field. We are looking at the biggest, deepest picture we can. And exactly to avoid these types of semantic arguments about academic dynamics, I am using the artificial manual metaphor to paint a picture with analogies (that aren't flawless either).

Well then, let's ask some big picture questions. Is thinking understood? Do we have convincing arguments that we are on the way to understanding it? Do we even have a good working definition of it? If we collect all the work that is done tied with cognitive science (including what's being done), put it next to each other, distill and polish the collection, do we end up with a single self-contained artificial manual for thinking, cognition, and the mind?[53] Or is it like a large multitude of manuals that can't be put together yet because either they are not saying the same thing (be it contradictory or speaking in different terms), or their contexts have no overlap and we can't see how they can sit next to each other let alone contradict one another?

---

[52] Choi, B. C. K. & Pak, A. W. P. Clin. Invest. Med. 29, 351–364 (2006).

[53] We are using the word thinking in a way that most cognitive scientists don't. Our usage is the broadest, not necessarily involving any agency(self), consciousness or even any macro-level brain functions. In this view a column of neurons or a single neuron could have some level or type of thinking.

Is it more like biology that there is a single hypothetical artificial manual for it except that it's just a mess? Lacking robust theories, in the sense that 1) it changes all too frequently (5-year shelf-life for scientific explanations being common), and 2) it's full of exceptions, effectively a manual that says in situation A it's like X, in situation B, like Y, and so on, where almost every instance is a new case. Biology may be a mess or full of too many exceptions but when it comes to independence and coherence as a field, it's a solid discipline.

Is it more like Physics that there is a single hypothetical manual for it, full of robust theories, but it's just not guaranteed that nature wouldn't have some surprises for us up its sleeve and render us wrong about some things later on? Well, surprises of nature can for sure rule out some cutting edge ideas or even change the fundamentals shrinking the domain of applicability of some theories, but it will never be the case that a well-established theory would not be applicable at all — there will still be some domain of applicability in which it's still perfectly accurate.

Is it more like anthropology that there is a single hypothetical manual for it, thanks to a solid axis of time and emergent physical processes to study objectively but that you have so little to go on that new evidence can change many things at every moment to the point that you may end up having to rewrite most of the artificial manual? That may be the case, but the past evidence would remain relevant and the associated analysis would just get recycled into the new version of the manual that replaces the old.

Neither of the two groups of cognitive scientists, those who believe cognitive revolution has failed, such as Núñez, et al. or those who oppose such statements like McShane et al. would defend a view that cognitive science is holding any single artificial manual, nothing like that of biology, physics, or anthropology. The situation for cognitive (and intelligence-related) sciences is that depending on one's interpretations, choice of semantics and categorization, either we have many different manuals or no manual at all.

If you go with many manuals then in too many cases it is not clear how two different ideas are compatible or contradictory. Take the example of "cognitive architectures ", if you're familiar with any, where rigorous comparisons between them can only be on the merits of the engineering principles behind them, not the cognitive science reasoning where each would have their own justifications. Each group or community of researchers would argue for its own. Don't mind trying to ever reconcile them.

If you go with saying there is no manual whatsoever, you would also have a point. Because cognitive science mainly amounts to either 1) a collection of clever observations, drawing from many disciplines, experiments, or philosopher's introspections, put together in some (reasonable and justified but fundamentally ad-hoc) framework and then described by either a set of English words that inherently lack scientific rigor or 2) some computational model that cannot hold its explanatory grip on mental phenomena, which then falls in the domain of artificial intelligence.

But why are we insisting on having a single artificial manual? Maybe cognitive science is just a different form of science. One that doesn't admit any imaginary artificial manual corresponding to it. Perhaps something like what McShane et al. call an "integrating science". We know that not only there are many different cognitive phenomena but also there are many aspects to each. There are many levels and layers to cognitive phenomena. It's never one isolated thing unlike the concept of a free particle in physics. There are many emergent processes and effects, some low-level (like how a neuron functions), some high-level (e.g. why and how you dream), all present simultaneously. Each layer of phenomena requires a different type of explanation perhaps even a different language for that explanation. And then there is the obvious variety in explanations of any given layer of mental phenomena that arises from the differences between the sets of what you initially hold true, and what aspects you are focusing your investigative attention on. So it seems reasonable to be satisfied with the absence of a single, coherent comprehensive manual. Maybe it's true, cognitive science is just different.

Here's where the missing manual problem metaphor comes in handy. Of course there are many layers and aspects to cognition. So maybe one could by zooming in say it makes sense to have multiple different manuals. But we know it's all one physical system, every part of it must work with every other part of it. We know that there are no natural boundaries (just like in biology) and there cannot be any internal inconsistencies. We know the missing manual problem is there. So by definition, there must exist a single artificial manual for it too. All levels of mental phenomena, high or low, have to have well-defined interfaces, well-defined couplings, and decouplings among some well-defined hierarchy of emergent phenomena. That is coherency. With it, you can think of multiple manuals as just one!

I believe this is an articulation of part of people's gut feelings during the cognitive revolution too. It's what fuels the conviction behind a "strong cognitive science" program. However, they thought "cognitivism" would take them there but it didn't happen. We are here as the new generation, inheriting their test results to do better. We know currently good work is being done regardless of how it's labeled but the problem is that we can't put them all together and see how to best evolve it.

You don't have to be a professional cognitive scientist to care about this. Per our discussion in the previous chapter, the missing manual of mind and thinking is humanity's problem. It's the manual we use to deal with any other missing manual problem, be it that of physics, social sciences, or any yet-to-emerge future discipline. So with that view, we ask again, is "thinking or cognition" understood?

Currently, if you ask this question from 10 different people (who have attempted to give an answer to it) you likely get 10 different lists with 10 different answers in each, where each answer, in turn, means different things to different people. This is by no means because of a lack of trying to give a coherent answer. Many individuals confidently feel they have a good satisfying working definition and nothing else is relevant at the moment. But here's the fact about the big picture: every decade, from the 60s till the present one (2010s/2020s), there are about 50,000 books published with at least one of the words, brain, mind, or intelligence present in the title. Never mind the number of articles or research papers on the topic which are an order of magnitude more.

So there is no shortage of claims of good explanations for cognition or thinking. Yet, if we want something fundamental, less context-dependent, less detail-dependent, more general, abstract, and rigid, with much fewer assumptions, we still have no clear starting point to go after it! This something is what the physicist David Deutsche beautifully summarizes in the phrase "hard-to-vary explanations". Therefore, in that sense we can finally conclude that thinking is NOT understood. Not even in the slightest, because we don't even know what we mean by understanding it.

Of course we have to bear in mind: the question of "what is thinking?" is not well-defined. It is a problematic one at least due to the complications of the usage in human language. Complications with the nature of the "what-question"[54], as well as the confusing correlation of thinking and cognition with the concept of "self" in us. We must note how much of all these are shaped by culture and not by scientific facts. We can certainly be sympathetic to that. We have been using the word "thinking" in this book in a way that most cognitive scientists don't. Our usage has been the broadest, not necessarily involving any agency (self), consciousness, or even any macro-level brain functions. In this view, a column of neurons or a single neuron could be considered to have some level or type of thinking.

---

[54] Think about the word "what" and the types of answers one could ever give to a "what-question", like "what something is", if there is actually a universally correct way to answer this question, without turning the question into another question like what something does or what it's made of. That is referred to as the nature of the "what-question", which comes with lots of philosophical complications.

Regardless of how we chose our semantics, in the giant collection of literature mentioned above, we are yet to find anyone stepping forward to say something like or analogous to: "we don't have the right foundations to give any satisfying answer to this question [of our thinking or equivalent]". It's either the case that people say we should totally abandon this question and replace it with some computer science problem, or that they are fully eager to provide answers which end up being not "good explanations", meaning they are easy-to-vary. People do indeed give their own variation on it, which results in the large collection of explanations in books and articles on the topic that fail to settle anything rigidly. The former case is of course nothing but the computationalism-only mentality. The latter case is much more worthy of contemplation, partly because professional cognitive scientists are fully aware of it already.

So, what do professional cognitive scientists say about this? To be concrete let's bring back the two groups we outlined and the representative case of Nunez et al. vs McShane et al. for each. They both agree that thinking or cognition is not understood. They both understand that it's still a young field. They just disagree on the path to get to understand it. The first group says we have failed on the path that wasn't going to take us there and now we are not sure about the future. The second group says "that was never a sure path but we're good now because given that cognitive science is in its infancy, if we continue we'll get to wherever we want to sooner or later". Both the attack and the counter-attack have some points we could agree on.

Our quest here is not to judge whether this current situation is good or bad but to ask, why is it this way? if it is a mess, why is it this particular mess? Why does this debate exist in the first place? etc. and not trusting that anything is trivial. In this view, it's too harsh to say the field has failed. Judging failure requires two things: a test and administering the test. McShane, et al. contests the failure claim on both fronts. Both on the judgment criterion and the measurements against the criterion.[55] Moreover, I don't think any intellectually-honest effort can be considered a failure.

On the other hand, McShane et al. and many others find perhaps too much comfort in the statement that the field is in its infancy. The "infant" metaphor is used to convey that we're demanding a level of maturity from the field that is not realistic so soon. Here's the problem. Blaming the incoherency on the age of the field is OK as long as we have some idea of how coherency and maturity are supposed to look like. And currently we do not! Looking at a human infant we may not know what they'd make of themselves when they grow up but we know what a mature

---

[55] Please see McShane et al. for details.

human being looks like. Strong cognitivism was one vision and the only. So in that sense, we should be sympathetic to Nunez et al. questioning the future of the field if that vision doesn't work.

We should also have sympathy for McShane et al.'s position because they are saying cognitive science is just doing what it can: there is currently no practical alternative for rolling up our sleeves and doing whatever can be done while maximizing creativity. Having said that, there is no reason to restrict ourselves to continue working inside the box of given philosophies of mind, science, and reality. And that's where we want to point the finger at. The silent and somewhat hidden philosophical impediments here.

As we saw, both groups raise very valid points nevertheless they find themselves in somewhat opposing views and neither group is fully satisfied. This kind of situation in a field can only come up when there are foundational problems. Both groups agree that they don't see what the mature future is supposed to look like. That's an agreement between them, even though one says the future is questionable and the other says there are no reasons it wouldn't be totally fine. When it comes to judging the maturity of a field of science it's very natural for any individual to think of physics. It's hard to argue with the maturity of physics as a scientific discipline. This is far too well-recognized already by cognitive scientists. We already mentioned the so-called physics-envy in the context of psychology, and it's just the same in cognitive science. Yet, if you ask cognitive scientists whether the mature future of the currently "infant" field looks like that of physics, they'd be quick to tell you "for sure not"! This is another agreement between the opposing views in cognitive science, albeit a silent one.

Where does this come from? If we don't know what maturity is supposed to look like, how can we know it will certainly not be like physics? This goes back to views that many have articulated in the past including especially the influential figure, Marvin Minsky. The view is very simple. It says physics is simple and the mind is anything but. While you can summarize the laws of nature in a handful of laws that fit on a T-shirt, the mind is far too complicated to ever admit such simplicity. Finding laws in Cognitive Science or Psychology is never going to be a fruitful route, not that Psychology hasn't tried that a bit already.

As you'd imagine, correct or incorrect, to say that physics is simple and the mind isn't, is quite a useless philosophy. The fundamental bias turns out to be the main challenge against forming the right foundational philosophies we need and the reason why we don't have them already. In Volume II of the book, we'll try to address these issues by introducing new philosophical views on science and the mind. What we have accomplished here is establishing the necessity of doing so. That is, to have a true science of cognition that could provide "good explanations" for "thinking and intelligence", we need to build new foundations or seriously augment the existing ones. That requires two major philosophical works. One for "cognitive" and one for "science".

In summary,

- Brain, mind, cognition, are all aspects of one system. It's one "machine", and it must have one manual!

- We have foundational problems when it comes to thinking and cognition. That is, we don't have a satisfying "outline" for such an artificial manual yet, nor can we say we are en route to getting one (regardless of how precise or imprecise the content).

- Arguments based on the fact that cognitive science is still an infant and it's too soon to understand something as complicated as thinking, are quite incomplete.

- In our review of the history of AI/cognitive science, we saw unjustified assumptions originating from our cognitive biases, most importantly what we introduced as the fundamental bias of our brain. We also argued that this bias plays its most acute role when the mind tries to establish a science of itself!

- On that basis, we proposed working on new philosophies, to form better foundations for a science of cognition, and by relation a science of intelligence.

Chapter 4

Foundations of AI as a Label

So far we have said that there is a missing manual of thinking. We looked into what science and philosophy have said about it, and we saw that the best place to search for the missing manual would be in cognitive science, and that we don't have a coherent artificial manual for it yet. Given that AI and cognitive science are intimately related, we covered the birth of both. However we did not get into the modern relation between them and the nature of their overlap. Instead, we moved on to a discussion only on the most recent state of cognitive science. In this chapter, we do the opposite and focus on AI, but we ask similar questions of it. That is, asking about the status of AI's artificial manual.

Separating AI and Cognitive Science

We start by making the distinction between AI and cognitive science clearer. Let's look at the overlap of the two fields first in terms of what AI brings to cognitive science. Even though we haven't yet formally defined AI we know it's about models that exist in computers. These models (and some of the theory behind them coming from computer science and statistics) can get recycled in studying a variety of cognitive phenomena. That kind of work is often labeled computational cognitive science or computational/theoretical neuroscience or computational psychology. These fields leverage the winning methods from applied mathematics, statistics, computer science, and similar disciplines to understand the human mind, and the brain.

There is a flow in the opposite direction too. Many researchers in AI look for biological plausibility both to get inspiration to develop a (slightly) new method, and to justify why some existing method is good. Therefore neuroscientific/psychological studies or observations can be useful both prior-to-development and post-development of methods in AI. The most basic example of prior-to-development would be the simplified model of neurons, the artificial neuron, and models for their activations.

Neuroscientific/psychological results can also be used post-development of AI methods, to give guiding justifications, say by providing weak forms of explanation for the goodness of the results, and why we should expect them. Famous examples in this category include giving a neuroscientific basis for convolutional neural networks or giving psychological basis to methods in representation learning and multi-task learning, for instance using the evidence in humans of reserving their attention to perceiving only changes in the environment that are relevant to the tasks at hand and not noticing other obvious changes.

So underneath the overlap, you can see two distinct pursuits: one is the engineering pursuit of building intelligent machines, and the other is a scientific pursuit of explaining human cognition and intelligence through computational modeling. The former is the field of AI and the latter constitutes the majority of modern work in cognitive science. Two separate fields, due to two distinct missing manuals. Similarly, we need to be mindful of the separation between the corresponding artificial manuals.

It is worth noting that sometimes the work in the context of cognitive science is in fact just a particular approach to AI, namely, the approach to building intelligent machines with the added hard constraint of having it operate very closely to how the human brain does, close in the sense of principles and methodologies. Researchers in this camp believe in the human brain (especially the brain of infants and children) as the ultimate guiding light and they try hard not to steer away from it, as much as possible given what we know about human cognition.

Turing's AI

So far we did not mention Alan Turing's name outside the context of Turing machines, and the foundations of computer science. He is widely considered the father of AI too. In 1950, he published a paper titled "Computing Machinery and Intelligence" in a philosophy journal, asking the following question right at the beginning: "Can machines think?" The title includes the word "intelligence", and the main question rests on the words "machine" and "think".

None of the three were properly defined back then, and his paper made it so that they wouldn't get properly defined for at least the next 70 years either! He argued that we don't need to define them because we can adopt an objective test instead. The test is playing a game which he named "The imitation game". There is a human judge sitting in one room exchanging written human language with another person or machine in a different room. If a judge communicating with a machine can't tell whether it's a person or machine on the other side, then the machine is said to be "intelligent" or considered that it's "thinking". This is the infamous Turing test.

Imagining such a test, Turing didn't see a need to differentiate between thinking and intelligence. Evident by the fact that the word intelligence appears only once in the entire paper, other than in the title. He also goes on to suggest that maybe we'll eventually find that the mind is all mechanical too, effectively discarding the need to define the word "machine" since even our minds could be considered a kind of machine, like any other mechanical machine people already knew of. We'll come back to this in volume II when we meditate on the word "artificial".

There are both practical and philosophical issues with this test such as, how to decide on who the judge is, how to administer the test, the scope, etc. For that reason some philosophers consider it too strong of a test for intelligence as it requires the machine, not being a human, to know so much about humans. Then there are others who consider it too easy because many people are easily fooled by a machine. As you may imagine, there are many variations of Turing's test, and even more attempts to pass it with many claims of having done so.

Nowadays most people disregard these philosophical issues and try to focus on the "spirit" of the test, a truly open-ended human-like conversation. Unfortunately, this "spirit" makes the test nothing but a circular definition, that is, to get to a clear definition of intelligence you have to define "human-like" conversation first which falls back on the concept of intelligence and vice versa. Nevertheless, it still serves as a defining guidepost for the field. The field that as we mentioned was labeled "Artificial Intelligence" by John McCarthy during the Dartmouth 1956 summer workshop. Nearly 6 years later after Turing's usage of the word intelligence.

Given the troubles with Turing's definition, the AI-dreamers said let's put the Turing test and his exact choice of words aside, let's focus on what he really wanted to tell us. Let's follow and build a discipline around that. But what DID he try to tell us exactly? Again it's like saying we should focus on the spirit of the Turing test. Under such lack of clarity, people's ideas of what the field should pursue naturally diverged. One simple idea was "look, it should be about all the tasks that right now takes a human to accomplish." There is no existing discipline with such a focus and it does need new scientific and engineering ideas and principles. It may receive ideas and contributions from nearby disciplines such as control theory but it (AI) stands to be a distinct pursuit.

Others tried to adopt a less human-based view for both practical and philosophical reasons. There are many things we desire to accomplish but no human can do so regardless of whether we consider it due to insufficient intelligence or not. Even in tasks that humans are well-accomplished, their performance is not strictly perfect. Humans just aren't perfect. And for AI we want something "perfect", more robust, and generally more reliable. That leads to a somewhat human-independent and idealized notion for the AI label. Which we'll get back to in a little bit.

What if we ask John McCarthy about it now? His most recent answer to the question of "what is AI?" (dates back to 2007), is as follows:

"Artificial intelligence is the science and engineering of making intelligent machines, especially intelligent computer programs. It is related to the similar task of using computers to understand human intelligence, but AI does not have to confine itself to methods that are biologically observable. And Intelligence is the computational part of the ability to achieve goals in the world"

That is fairly in line with what we just said. Although you must notice the circular usage of phrases such as intelligent machines or programs. However, one can treat these as mere labels. But still, labels have to be put on something. That something should be a *Formal* concept (formal means based on set axioms and therefore non-circular).

The Bible of AI

Stuart Russell and Peter Norvig, two well-known computer scientists, in a monumental effort wrote the most successful and most widely adopted textbook for AI. It's titled "Artificial Intelligence, A Modern Approach". Some people like to refer to it as "the Bible of AI". It is the main go-to reference on anything basic or introductory in AI. Anything you may have seen around on defining AI, most likely comes straight from their text. Russell and Norvig, quite appropriately begin the book by defining the AI label. So it's important for us to briefly review their discussion first.

By separating human-like and human-independent, and separating thinking from acting, they chart 4 types of possible meanings (2 factors each with 2 possibilities) for AI . Systems that can 1) think humanly 2) act humanly 3) think not-necessarily humanly but rather perfectly 4) act not-necessarily humanly but rather perfectly.

They argue that we should stick with the last category i.e. "acting perfectly". Informally, we know that removing humans from the definition allows for idealization. Also, given that thinking isn't understood yet, per our previous discussion, sticking with acting and behavior instead of thinking, allows for a mathematical formalization. Therefore, "acting perfectly" looks promising as a candidate for an ideal formal definition.

Literal perfection is not a realistic concept, so to formalize we toss the word *perfect* in favor of *rational*. Russell and Norvig highlight the difference as follows: "Rationality maximizes expected performance, while perfection maximizes actual performance." As such, they define the label AI as the pursuit of building rational agents.

Let's unpack that a bit. The word agent stems from the Latin word "agere", which means "to do". That captures the *acting* part of our definition, whereas "acting perfectly" gets formalized by the concept of a rational agent: "A rational agent is one that acts so as to achieve the best outcome or, when there is uncertainty, the best expected outcome"[56]. The best expected means the best average if you could try performing the task over and over again indefinitely. This is the main idea behind utility-based decision theory which combines utility theory with probability theory to "average" over all the possible outcomes of actions.

Utility theory, the basis of this definition, assigns some quantitative utility value for any outcome based on a set of preferences that satisfy some axioms. Thus the goal of every rational agent is to maximize some expected utility. And acting otherwise, given what you know, would be an irrational act under this definition.

Rational agents provide the best definition among the 4 categories under comparison by Russell and Norvig. They succinctly articulate the benefits of this choice, in the following quote:

"The rational-agent approach has two advantages over the other approaches. First, it is more general than the "laws of thought" approach because correct inference is just one of several possible mechanisms for achieving rationality.[57] Second, it is more amenable to scientific development than

---

[56] More formally: "For each possible percept sequence, a rational agent should select an action that is expected to maximize its performance measure, given the evidence provided by the percept sequence and whatever built-in knowledge the agent has." Russell and Norvig, 3rd Edition.

[57] "Making correct inferences is sometimes part of being a rational agent, because one way to act rationally is to reason logically to the conclusion that a given action will achieve one's goals and then to act on that conclusion. On the other hand, correct inference is not all of rationality; in some situations, there is no

are approaches based on human behavior or human thought. The standard of rationality is mathematically well defined and completely general, and can be "unpacked" to generate agent designs that provably achieve it."

Having mentioned this, given that thinking is not well-defined, one may object that: the assumption of thinking being separate from acting, is not entirely rigorous, unless we insist on thinking being defined precisely such that it would involve no acting, as we almost always do. However, we can also always think of thinking as an action that results in some change in some environment (e.g. wherever the computation is happening in, physically speaking, or the mental-state-environment of the agent that is perturbed by the agent's choice to "think"), whether or not it results in any logical statement/conclusion or valid inference. This can get resolved in the following way: 1) The environment where the change due to thinking occurs in, may be considered not the agent's environment, that is, not connected to the reality that the agent experiences or expected to perform in. 2) If the two environments (where "thinking" occurs and where the agent performs actions in) have consequential overlaps, then we simply consider that thinking as a form of acting subject to the same demand for rationality. So with that, we can restore rigor to a practical definition for AI as a rational agent.

Rethinking Rational Agency

Are we done? Do "rational agents" provide the final definition for the AI label? Well, not so fast. Of course, just like Russell and Norvig argue, it's a good move to come up with a definition that is decoupled from human psychology. There remains a couple of issues however. First, there are other factors that we could think of to create more than 4 categories and to isolate a good target for AI within them. Factors such as generality of methods, efficiency, the elegance of conduct, swiftness, lower need for maintenance while being beneficial, on and on. However, we can work around that. Let's assume all such factors would just serve some ultimate purpose, and further assume that the ultimate purpose is captured by the goal of a rational agent.

Second, there are too many things that could fit the bill of "acting rationally", yet we don't want to call them AI. For instance, any computer program is a rational agent except it just doesn't know

---

provably correct thing to do, but something must still be done. There are also ways of acting rationally that cannot be said to involve inference. For example, recoiling from a hot stove is a reflex action that is usually more successful than a slower action taken after careful deliberation." Russell and Norvig.

much about anything. We can somewhat resolve this issue too. In two ways: 1) by requiring the rational agents of AI, to exhibit some degree of operational autonomy, environmental perception, adaptation to change, goal-orientation, etc. 2) by demanding that the goal of the rational agent should possess more than some minimum degree of complexity, which would force the agent to be sophisticated enough to be worthy of the label AI.

As you can see, both ways include subjective elements and therefore not entirely satisfactory. The philosophical challenges around the words "agent" and "autonomous" are severely downplayed in their usage in AI. The concept of agency is not a well-understood one. What kind of things we want to call agents, is totally tied to our own human experience, it is where the fundamental bias shows up again.

Third, although the concept of rationality introduced in the last section may seem like an innocent one and backed by utility theory, it too suffers a great deal from both philosophical and mathematical complications. Let's go back to the basis of it, which is mainly in the John von Neumann and Oskar Morgenstern utility theory (1947). The theory starts with the assumption that there are a known set of choices the agent can choose from, and the agent has some preferences over those choices. Furthermore, those preferences must obey a set of axioms (these are the axioms of utility theory, mathematical properties that the preferences must satisfy). Under these conditions, von Neumann and Morgenstern prove there must exist a utility function, such that a set of rational preferences would maximize its expected value (the average of many trials). Think of utility as the scale on which preference is measured. Put your preferences on the utility-scale, and it shows how much is the average weight/value of your choices. This usage of utility theory in decision theory requires what's known as the expected-utility hypothesis. It seems natural, why is it a hypothesis?

In reality, we (humans) often start with what we want, the goal, and work backward to figure out what our choices and preferences should be. It doesn't mean we are aware of all our goals and actually calculate our preferences. Nevertheless, to an economist, we can exhibit a behavior as if we have calculated them and are aware of our goals. In that case, the data they collect on us can be described accurately with utility theory, and they can deduce our preference sets from it. Let's call these utility-theoretic deduced preferences, the set of pseudo-preferences as opposed to actual preferences of a human. Why would they be different?

If an observer tries to use an apparent utility function deduced from the behavior of an agent, the results could be interpreted as pseudo-preferences of the agent with no guarantee that they would match the agent's true set of preferences, at least in size and order of the preferences. Just like the true desired goal of the agent may be different than the deduced one.

The proponents of rationality can argue that for a rational agent compatible with utility-theoretic assumptions, any such pseudo-preferences would coincide with true preferences, up to some transformations that don't change the ordering of preferences. By the definition of rationality based on utility theory, that is correct, because utility theory is self-consistent and does describe a certain class of systems, namely those that satisfy its (pretty general) axioms.[58] The question is: what are these systems (as general as they may be) and how well do they apply to our real world of humans? Or are we just going to call humans irrational anytime they cannot be modeled by utility theory given how general and reasonable this theory is?

Within cognitive science and economics, the concept of rationality translates to what's known as "rational choice theory" and it purports to withstand challenges from all psychological philosophies. Yet as one can imagine, there are many challenges with the notion of rationality applied to the real complex world of humans. For instance, in game theory, where rationality is applied to strategic settings, if one insists on having strictly rational or near rational agents many paradoxical or undesirable outcomes follow.[59] These observations together with other experimental findings led Daniel Kahneman and Amos Tversky to develop what's known as Prospect theory in economics. It is founded on a totally different philosophy than utility theory. The starting point is to forget about rationality and instead formulate a theory to predict the actual behavior of human agents. This work won the 2002 Nobel prize in economics and set off the whole vibrant field of behavioral economics, where experimental methods are used to build economics theories.

Humans are observed to abundantly engage in irrational acts according to rational choice theory. This does not mean irrational in the literal sense of the word, rather in the sense that we may not necessarily act in our own best *apparent* interest, or we may have conflicting preferences that are not compatible with at least the axioms of utility theory. Examples in consumers are well-documented where we assume we know their goals by definition, say getting better products,

---

[58] For instance, the fact that we should use only one scale to measure the set of preferences and each preference would fall in exactly one point on this axis. So if you prefer A to B and B to C, you must also prefer A to C, also (axiom of transitivity).

[59] Bratman, M. (1992). Planning and the stability of intention. Minds and Machines 2: 1–16.

Kreps, D. M., and R. Wilson. (1982). Reputation and imperfect information. Journal of Economic Theory 27: 253–279

Selten, R. (1978). The chain store paradox. *Theory and Decision* 9: 127–159.

cheaper, faster, safer, and so on.[60] Cognitive science also documents systematic errors in reasoning by humans.[61] A proponent of rationality, as a good candidate for the AI-label, may say "yes utility theory runs into problems for human agents but in AI we are considering ideal agents immune to psychological complications of humans".

Here's the key question that has gone utterly overlooked. Yes, we control AI agents and we can try to stay away from such complications, we can pursue building "rational agents", but what about the environment the agent operates in? Can that be controlled? Recall that we are pursuing "rational agency" as the right candidate for Alan Turing's version of AI which ought to act in the *real world of humans,* to be of any use.

In almost any theoretical work, we heavily simplify our target environment (either explicitly or implicitly). We consider a compact and abstract slice of its reality that we believe is most relevant. That results in what we shall call an *artificial world*. We have a lot of ground to cover before we can properly define the word artificial and the phrase artificial worlds, in volume II. But for now, take it to be an abstract slice of a tiny patch of all simulatable reality. Or roughly a simulated world. For instance, in the early days of AI, there were demonstrations of successful programs accomplishing tasks in limited domains, which went by the name microworlds. The most famous microworld was the Blocks world. A typical task in this world would be to rearrange a set of simulated solid blocks placed on a simulated tabletop in a certain way, where the agent is a hand and can pick up one block at a time. Nowadays we have AI programs deployed for real-world applications, so they are operating in much bigger domains and it may not be appropriate to call it microworld, but still the system can be said to be living in an artificial world since it operates based on heavy approximations to the real world.

One may say, "it doesn't matter what world an AI lives in, its actions are actuated in our real world. Those are what we aim to judge them with, so they can be said to live in the same world as us for all practical purposes." Well, that reasoning is nothing new nor particular to AI. Within operations research, embedded systems, and stochastic optimal control, people have always been doing the same thing; define sufficiently complex approximations of the environment and consider some appropriate set of (environment) state variables accordingly. This can be interpreted as some microworld approximation to the real world in which the system is going to operate.

---

[60] Dan Ariely, "Predictably Irrational" (2008).
[61] Kahneman et al. Judgment under uncertainty: Heuristics and biases. (1982).

In all these cases one can still work to put provably good bounds on the errors induced by the approximations to reality. But note that such disciplined engineering can never be perfect, rather just measurably good enough such that it gives us the confidence to deploy them in real-world situations. That is far from the perfection needed for the degree of autonomy that the AI-label seeks to be able to represent. In other words, the rational agents are only rational in the artificial world. For rational agency to be a meaningful candidate for the AI label, it needs to be a concept in the real world. That is how can an agent be called rational in the real world? The right question to ask therefore becomes: to what extent could or does the world that an AI system lives in match our real world? Is approximating the real world ever going to be sufficient for the concept of (utility-theoretic) rationality to survive beyond a notion that is only fully applicable to artificial worlds?

Before going down this road, let's remember what we are trying to earn. A utility-theoretic rational agent is just a formalization of what we want the AI-label to represent. An agent that always does the "right" thing for the ultimate maximization of expected utility given its resources and what it knows at any given time.[62] In other words, an agent that behaves as ideally as possible. So we may ask: can we somehow bypass this "artificial world" complication and fix/extend the notion of rational agent to define the AI-label?

Well, I hope by this point you can see that the concepts of agency, rationality, and goals (a major topic in volume II) are in fact what we directly borrow from our human experience and human realities. They are not as human-independent concepts as proponents of "rational agency" would like them to be. Perhaps, trying to separate the definition from humans and their reality may not after all be the way to go. The concept of rationality seems like something we understand but further examination reveals how shaky it is. Similar to the word knowledge, we think we know things but when we examine it further we get stuck in epistemology — the study of what knowledge is. Or just like when I say I, or I am. Examine it deeper and you get stuck in ontology — the study of what being is. Oftentimes the simplest things to state or communicate to another human are the hardest ones to define rigorously.

As I mentioned, prospect theory in economics gives up on the rationality of humans because utility theory fails to explain many behavioral phenomena, precisely those that prospect theory aims

---

[62] I am borrowing the terminology of "do the right thing" from Stuart Russell and Eric Wefald, authors of the AI series book titled just that, "Do the Right Thing", published in 1991 containing compilation of their work in rationality given limited resources, i.e. limited rationality.

to explain. However, such "effective" theories as useful as they may be in economics, are hardly useful for understanding the actual agent. What I mean is that they don't explain WHY humans act the way they do as opposed to many other ways one could imagine them acting. What would be desirable to see is a theory that ventures out to restore some rationality to the existence of human irrationality. That theory would by necessity be at least a parent of prospect theory and perhaps would replace rationality with something that is not mindless of physical reality!

Both utility theory and prospect theory are built presumptuously on agents with goals. Yet, we don't quite know the true goal of the physical systems representing humans in these settings under study. This statement may be hard to digest, may even sound a bit demeaning to human agency, and we won't expand on that just yet. But consider that the word "goal" is another one of those words that we think we have a good handle on and upon further examination it all comes tumbling down. For instance, consider that every goal is a subgoal of another. Or that there is no such thing as goals in the fundamental laws of physics. So what are goals really? Of course we could put philosophy aside and get practical, but then we should put the meaninglessness of the labels aside too and accept that the only meaning of the label AI comes from what people actually do.

So what DO people actually do? We build agents while having in mind a task or a set of tasks they should perform in some environment. For instance, the task could be to recognize my voice from that of my colleagues in any place at any time. The utility we want the agent to maximize (the only utility we really care about) is one that we cannot directly specify. That is, the *true* utility is only an intention of the engineer who builds the agent. So let's call this true utility, the *intended* utility. Question is, how can the agent be judged as rational or irrational if it doesn't know the utility it must maximize the expectation of?

Let's say it can't. But the engineer can still try to enhance the performance of the agent in service of the intended utility. First, one has to choose a set of tests designed to approximate the measurement of intended utility maximization. That is, we choose to judge their performance based on how well or how much they do according to what we know to be the right thing to do in our world. Next, some utility function for the agent must be designed such that upon its maximization, its performance on the representative intended utility (the pre-designed tests) increases. So we indirectly maximize the intended utility (the desire of the engineer) by adjusting the utility of the agent.

The agent's actual utility is a complicated thing to analyze. Most people in machine learning may think of actual utility as an objective function (or a loss function) and that would be only partially correct. It's true that the agent's actual utility is mostly defined by this function (and its structure) but almost everything else in building an AI agent also affects what its actual utility ends up being. The data and examples it has seen, their representations and the model, training process, the rules and prior knowledge it's supplemented with if any, etc. all affect its actual utility, which is something emergent out of the whole system. This actual utility is something that the theoretical work in machine learning tries to figure out. For instance, in deep neural networks one tries to figure out what the network is "really" doing, not just how it was built or what loss function was used. The goal is to make the system such that its actual utility becomes passing the tests better. That is, to be closer to the intended (true) utility.

Since the intended-utility is being indirectly maximized, maybe we can define a notion of "indirect rationality". That would mean to define the AI label as one that aims to build "indirectly rational" agents that maximize the expected utility that its designer intends to maximize. Suppose, there is a superstar engineer who has done this job perfectly, and now the agent is performing perfectly on what it is being tested on. Can the agent now be called (indirectly) rational? Before jumping to any answer here, first recall that as we said, only *perfection* is concerned with the actual performance. Rationality on the other hand is NOT judged by actual performance/outcomes.

Now, in the case that the actual performance came out to be perfect thanks to the perfect engineer, how could we say that the agent didn't do the "right" thing and isn't rational? The answer is simple. It's true that it performed perfectly on some sets of tests, we still don't know with certainty how it would perform on a different set of tests. The more it performs well, the higher our expectation of observing only rational actions from the agent. Yet, there will always be a chance it would violate all our expectations and act irrationally. In other words, indirect rationality is ill-defined because the intended-utility is not an actual utility function we can specify.

Consider the game of Go, where AlphaZero (Google Deepmind's GO playing program) is currently a winning champion. The intended utility in this case is winning the game against any possible opponent. Being a world champion is not proof that AlphaZero is not making mistakes. A yet-to-come superior algorithm that could beat AlphaZero could prove the irrationality of AlphaZero just by winning. So the answer is no, we can't even call it "indirectly rational".

One may fairly object to this conclusion saying what if the set of possible tests were finite. For instance, what if the intended utility (the intention of the engineer) was to just beat Lee Sedol, the former GO champion of the world, and nothing else? Surely in this case indirect rationality is satisfied but this would be a very limited, albeit practical, concept of rationality, and certainly not what the AI label is seeking to define itself with. In fact, most important applications of AI and machine learning would be in uncertain domains with an indefinite number of possibilities to choose tests from.

Another also fair objection to this conclusion is as follows: "Yes, strictly speaking, indirect rationality is not a realistic concept, but maybe that is too high of a bar, what about approximately (indirectly) rational? Let's clarify what we are approximating first. Suppose we are approximating the set of possible tests. Suppose we know something about its statistical distribution and we are sampling from it such that it lets us be sufficiently confident about the quantification of errors we are making. That would be quantifying confidence on some measure of the difference between the actual utility and the true. This is where the notion of a "probably-approximately-correct"(PAC) learning comes in, which we will thoroughly review later in the context of machine learning. We will see that there are a few assumptions we must accept in order to come up with a bound on the statistical risk, such as assuming the distribution where the tests are going to come from remains unchanged.

In the real world where AI wants to be, these guarantees or bounds may not be satisfactory or useful in all cases and operating conditions, such as when the amounts of data available are not sufficient, or when the engineer's choices for what approximation and what confidence on guaranteeing those approximations are sufficient, may not reflect the actual intended ultimate use. In practice, this is dealt with only by increasing the number of experiments until the statistics are within the risk appetite of a committee of humans.

What we have just described is what I call the artificial version of Hume's induction problem, or simply the *artificial induction problem*, a statement that says we cannot have true indirect rationality. David Hume observed that induction, inferring a statement based on a set of observations more general than the observations themselves, such as coming up with a natural law not just for the past but also the future, is not engaging in an act of reasoning. That is, it doesn't come with any guarantee. This is a fundamental problem with induction. Indirect rationality (artificial induction problem) is a problem exactly in the sense of Hume's induction problem, except that in the case of artificial induction there is the added part that we make claims over outcomes coming out of an artificial world, where those outcomes are chosen by an emergent behavior of a system we have built having in mind some intended utility that is being indirectly optimized by how we build the system. In very simple cases like in traditional AI with logic and analytically-understood algorithms, we can guarantee the behavior solely in service of the intended utility. But such programs are far from those in modern AI as we will see. In fact, we don't need AI for tasks that we can analytically specify.

Alright, what if we ignore all these complications and assume we could hypothetically form an agent utility that would coincide with the intended utility? Could the concept of rationality be demanded perfectly now? Still NO. The way that utility-theoretic rationality is defined, i.e. "do the right things given what you know" has some degree of perfectness to it that is not realizable and therefore non-realistic. That is, it ignores the finiteness of computation and memory resources. Even in an artificial world things can't be perfect; they take time (compute time) and space (memory space). Therefore one must bring in notions of time and space complexity (concepts in computer science to quantify the compute and memory requirements of an algorithm relative to its input size) into the definition of rational agency.

Computer scientists have tried that avenue to some degree. For instance, one proposed idea is *calculative rationality* whereby rationality is judged by the presence of a calculation that would have eventually found the "right" thing to do given enough time in between actions. Maybe if AlphaZero had infinite computing power and infinite memory one could prove that it would truly be the ultimate champion, or strictly indirectly rational.

This is of course an unsatisfactory definition, just as a good choice that cannot be taken is. A better candidate is known as *bounded optimality*. The idea is to take into account the finiteness of resources of the agent and condition its rationality on that, similar to the concept of bounded rationality in humans introduced by Herbert Simon's work that won the 1978 Nobel prize in economics.[63] Rationality of bounded optimal agents is judged based on the optimality of their programs (which must take into account resource constraints) rather than their actions. This sounds like a good concept to follow up on, however, it's theoretically an infant concept relative to the role rationality wants to take over in defining the AI label.

So far, the question was whether we could meaningfully call any agent rational. Now, let's discuss whether we can call any agent *irrational*? Suppose the agent is not engineered perfectly as judged by the intended utility and thus it makes mistakes. In reality, this is going to be almost always the case.

---

[63] He suggested that given their finite computational resources, humans are severely time-bound and deliberate only long enough to come up with some "good enough" answer; an act that he labeled *satisficing*.

Consider the example of a self-driving car where its intended utility is to be a "perfect driver"; no accidents ever; no disturbances to other drivers or objects around; keep everyone happy to the extent possible by a car. We know by definition it's not going to be able to fulfill that. But what's worse is that it's still going to crash into objects from time to time that no human driver would soberly. Why? Because not only the actual utility of the car doesn't completely match the intended utility by our assumption of imperfect engineering but the actual utility of the self-driving car is not even about driving or not crashing into objects. The actual utility that it ends up adopting has nothing directly to do with humans and their world. It lives in its artificial world per our discussion above of how the actual utility emerges.

Ignoring resource constraints, that actual emergent utility could be said to be acted on perfectly rationally. So how can the car be said to be irrational? Well it can't, it's only our own irrationality that causes us to call the machine irrational. It's the fundamental bias of our brain that defaults us to use the lens of human realities to judge the machine by. In software engineering, coding errors are called bugs in the software. Does the bug really exist in the software? No, there are no bugs, there are no errors in programs. The only errors are in programming, namely with the human, where errors are defined by the difference between the rational program/machine and the intentions of the engineer for the machine that only exists in the head of the developers (and hopefully their software specification document).

If we remove ourselves from the picture, we see that all artificial agents are always rational in their own artificial world. We can call this *direct rationality*. Direct rationality is a perfectly valid concept for either when a machine judges its own rationality or when a human judges its[64] own rationality. Indeed, the only requirement is for the judge of rationality to share the same reality as the actor, that is, to be the same agent. To drive that last point home, consider one person (the judge) passes the irrationality judgement on another (the actor). This can only be valid if the judge is aware of the goal of the actor — not necessarily what the actor says is his or her goal. And that is not possible for the judge if you believe that the goal is represented by the entirety of the physical system representing the actor which the judge cannot readout, at least no technology can do so yet! That is yet another example of indirect rationality failing as a concept.

---

[64] Since knowledge of the goal is needed and we can't assume it's known by the conscious self, here "it" refers to the human machine, the entire physical system it can be said to represent.

Now let's go back to artificial worlds. Since we can't directly judge the rationality of artificial agents, can we build an observer AI agent that judges the rationality of another AI on our behalf? That would be indirect-indirect-rationality as it involves an extra indirection.

Suppose an observer algorithm is completely unaware of all the engineering details of a self-driving car including the fact that it's not being driven by a human, except that this car is supposed to be a rational agent just like the observer algorithm itself is supposed to be too. By observing the behavior of the car as a black-box, the observer tries to infer its goal. Now suppose the car ends up acting in a way that the observer can best explain by road-rage and intentional maneuvering to cause an accident for another vehicle. Wise human observers may attribute that to stupidity or irrationality than to malice, had the behavior come from another human. However, we still don't know how to define irrationality well enough to be able to let the machine learn about it too. Therefore, instead of inferring indirect irrationality of the agent, the observer may just infer a utility/goal for the agent that is quite different from the intended utility that the creator of the car had in mind.

Building machines based on idealized notions of rationality is clearly missing something. This observer-dependence once again signifies how much more subtle the concept of goals and rationality are both in humans and machines than what a naive utility-theoretic mindset may suggest.

To stretch this last point further, note that we totally glossed over what the intended (true) utility is supposed to be. Namely, where do these utilities come from? Utility-theoretic rationality has absolutely nothing to say about that. Earlier we said it's just the intention of the engineer, but what is the intention of the engineer? In our world, it's whatever the intention of the entity which the engineer is working for — most likely sustainable financial growth subject to some constraints. It is not even the intention of any single person but perhaps that of an emergent socio-economic structure out of many humans and their belief systems morphed into something unrecognizable. Think about the intended (true) utility behind for instance YouTube or Google search algorithms versus their actual utility. Keeping in mind that their actual utility ends up shaping the external environment (it's users behaviors) whereas their intended (true) utility considers them to be independent of and only in service of the external environment. What if we need to demand from the candidate definition of the AI label to have something to say about what the utility should be? After all, no human seems to get their intended utility from another human, at least not fully.

The intention with this section was only to expose some shortcomings of rational agency as it's currently understood. Now in sympathy for it, rational agency does seem like an innocent concept, meaning it's reasonable to assume that we should be able to pinpoint it and lock it down. That's because we all have seen examples of "bad thinking" either in ourselves or in others, all cases that seem totally correctable. For instance; when we have clear long term goals but we make decisions based on only short-term impulses; when we don't incorporate all the knowledge we do have for better decision making; or when we don't leverage the fact that the effort to acquire some new relevant knowledge is less that the loss due to poor decisions due to incomplete knowledge. So it definitely seems like we can and should formalize some ideal notion of rationality that is free of these kinds of mistakes, at least to the extent possible given the available resources to a person/agent.

Unfortunately, as we saw, in its present formulation, that is when we jump to ideal notions (of rational agency), we run into many philosophical issues that suggest maybe we've jumped too far and stumbled upon something that has no grounding in reality. We can neither make definitive statements about the rationality of agents nor their irrationality that are universally true. Universal here means valid as a notion to serve as a definition for the AI label. This does not mean that rational agency cannot serve as good inspiration. In fact, it has served as a potent inspiration throughout the entire history of the AI field.

The usage of the notion of rationality is obviously not limited to building machines, we humans use it all the time on judging one another too. Coming from our human experience about our own decision making, especially in life-long-type planning decisions, we can spot unwise choices in usually younger folks and we are often too eager to call them out as "thinking mistakes". Yet, as we argued, no currently known notion of rationality holds water universally. The challenges we are facing in formalizing rationality are yet another manifestation of the same issues we faced in defining thinking. These problems all lie at the core of the broader missing manual problem which we stated in chapter 2. In particular, you may have realized that defining rationality runs into many problems that defining "better thinking" does too.

In conclusion, AI remains to be just a label. A label that is sometimes put on the box that contains what people actually do, inspired by the unrealistic and idealized notion of rational agency. Although words like rationality and intelligence are related to what's being done inside the box, they can't legitimately be used as labels since they still lack proper definitions. We use the concept of rationality on judging one another in our everyday human world successfully, only because our brains experience enough shared reality to allow us to get away with the approximation we introduce in doing so.

Reasoning and Logic

We just tried defining rationality based on the choice of actions an agent makes in service of its goal. We didn't talk about how. The process to arrive at those choices or decisions is typically called reasoning. However, consider an agent that invents a random process to choose among available actions, does anybody consider that a reasoning process? No, because we all have some informal intuition about what reasoning is supposed to be, so how do we define it? What *IS* reasoning? Should we call it reasoning only if the agent arrives at the "right" or "best" possible choice/decision? Maybe not, because that would be too high of a bar, we can call that "rational reasoning", following our discussion in the previous section. In fact, reasoning can be of much lower quality and we know we'd still call it reasoning but how much lower? We already said it, namely, random choice making is the lower bound. [65]

> *Reasoning is any process towards making a choice that isn't random.*

Therefore, *reason* can be defined as anything you can express in any way that shows your choice isn't random. Reason is tied with un-randomness. A reasoning process is an attempt to help the agent do better than random. I say "attempt", because the outcomes of many poor reasoning processes especially in uncertain and complex environments can end up much worse than that of a random process. In gambling for instance, if your reasoning is based on a specific process with a systematic error, you may do extremely worse than a set of random choices. By this token, in reasoning, we're mainly focused on the process of arriving at the choice of action rather than the outcome of the action. Though, we may call it poor or irrational reasoning if it results in bad outcomes, in accordance with what we just referred to as "rational reasoning".

It seems like our definition depends on the concept of random. But how can we prove something is random or nonrandom? This is in fact among the most fundamental questions you can ask of nature and of course people have been at it forever. Trying to answer this question gets

---

[65] Note that just like in the previous section, choices and actions are pretty broad terms in our discussion. An example of a choice of an action could be a decision you make to believe in something in some way e.g. as true, real, or useful.

to the heart of the foundations of probability theory and randomness. In short, once you make certain assumptions about randomness, then you can easily define some tests of randomness. Still, saying that something is truly random provokes open philosophical problems.

Andrey Kolmogorov, the famous soviet mathematician which we will have to talk a lot more about in volume II, got closest to addressing these questions. What we want to note for now is that there is a fundamental link between randomness and reasoning. If one is not fully understood, so isn't the other, and you may very well consider both as part of the missing manual problem. Concepts of non-randomness and reasons are tied to each other at the hip, which may suggest that ultimately a mere reasoning process may not be able to fully solve the issue of whether something is truly random or not without referring to circular arguments.

What do you mean by "mere reasoning process"? Let us, for now, set aside both the issue of the goodness of resulting outcomes and fundamental issues around randomness. Obviously a reasoning process can be simple, and it can be very complex. Can we say anything about the correctness of a reasoning process independent of how sophisticated it is? We have all been in situations in which there are a lot of unknowns but we still want to make sure we are reasoning properly about it irrespective of how much we don't know. In other words, we want to make sure that at least our processing of what we do know is correct. Does this ring a bell? We even have our folk terminology for referring to false reasoning processes or when people present false arguments. Yes, we call them illogical. We want our reasoning to be logical even if it is insufficient or incomplete. We often have a gut feeling about it when we aren't being logical.

Well, so did the ancient Greek philosophers, particularly and most notably Aristotle, who wanted the spoken words to be *sound,* to be correct. He called it analytics in his book titled "Prior Analytics". To Aristotle, the correct *process* of reasoning should be a fact on its own that is *a priori* true regardless of what you want to reason about. So for him, this was "Prior Analytics" as opposed to "Posterior Analytics", the title of his next book written mostly about definitions and what we may consider scientific knowledge. So then, Analytics was the branch of his studies concerned with *analyzing the correctness* of what people say and the statements they make.

The word "logic" is the transformed version of the Greek word "logos" which originally meant "what is spoken". Roughly 2300 years later, there is still no universal consensus on the exact definition and boundaries of logic. What everyone agrees with is that logic is about the systematic study of the correctness of various forms of inference. That is a bit circular because defining inference falls back on defining reasoning which in turn brings up logic. However, we can get around that by separating knowledge from the reasoning process itself. Even though you cannot reason without any knowledge, we can separate the process from what is being processed. Inference is a much broader concept because it involves at least both reasoning AND knowledge.

Having established that, we can now make a choice of definition for LOGIC, consistent with everything we have talked about. That is, logic is the study of CORRECT FORMS of reasoning. When your form of reasoning is correct, we can call it logical reasoning. Note that this is quite distinct from rationality and rational reasoning which requires goals and results much beyond just a correct FORM of reasoning.

OK. We have a definition now and it's quite broad. Just as there are different forms, representations, and contexts for reasoning, there are many branches of logic too. First off, we have our own everyday folk logic, working straight in the domain of human/natural language.[66] This is called informal logic as opposed to formal logic in which expressions (or *sentences*, technically speaking) are all crystal clear to anyone who knows all the (formal) rules of that logic.[67] That means sentences in formal logic do not require to be accompanied with any human intuition, and therefore machines can verify and operate on them too. For that reason, formal logic must work with symbols. In other words, formal logic is a special case of symbolic logic.[68]

Symbolic logic may still include some ambiguities that require a human to be around to work with them. For instance, all mathematical logics are symbolic but they are not all necessarily formal. When they are not, they are intended for only mathematicians to communicate with each other. In many cases, it's still too hard to formalize them such that a computer could express the same content. Therefore even though it'd be nice if we had a formal version of all logic, the requirements of formality are still too high in some application domains and contexts where we still can't quite free ourselves from our culture and human language. So it remains a work in progress to be able to "formalize" all the things we may find useful to talk about in informal logic. That's why

---

[66] "Natural language" is a name that computer scientists have chosen for our culture-based human languages in order to distinguish it from programming languages and machine language.

[67] This includes syntax (rules for what are valid sentences) and semantics (rules for what are the truth values of the sentences).

[68] Symbols are anything with only formal content. For instance, words in our everyday language can be considered atomic symbols, in the sense that "at" is not a meaningful part of "atomic", it has no value on its own. The word "atomic" is just one atomic symbol.

philosophical logic (another branch of logic) despite its tremendous overlap with mathematical logic is still considered distinct from it.[69]

Aristotle was after a formal logic representing anything that one could speak of. That is almost an impossible task which logicians are still trying to achieve. Aristotle himself made a lot of progress. The main form of reasoning that he solidly introduced to us (in prior analytics) is known as syllogism. Syllogism is a specific kind of deductive reasoning (sometimes referred to as "term logic"). Aristotle's syllogism is the most famous kind of term logic in which he combines a general premise with a specific premise to deduce a conclusion. Here's an example:

- General Premise: No computer program lives in our reality.
- Specific Premise: AlphaGO is a computer program.
- Conclusion: AlphaGO does not live in our reality.

You notice that there is a pattern in this reasoning. This pattern is what's known as the "argument form". The argument form in the above syllogism is:

- No A is in B.
- C is an A.
- Therefore, C is not in B.

This is a correct form of reasoning, i.e. it is logical. However, if for instance you don't believe that the general premise in the above example is true, then my argument is not *sound* to you even though it is *valid* because it has a correct syllogistic argument form. As you may imagine there are many other patterns and reasoning templates that we may want to consider valid or logical that fall outside of syllogism.

Since we are concerned with AI here, let us narrow our definition of logic to what has been relevant to AI which may not stay the same in the future. We said, logic is about correct forms of reasoning, and we can make that a bit more specific by including what kind of knowledge is being processed in logic. In the world of AI, when people say logic they refer to correct forms of reasoning with definite knowledge. By definite knowledge, we mean something that is either true or false. For

---

[69] In fact, most branches of formal logic in mathematics were initially spun out of the work in philosophy. A great example here would be *modal* logic that is increasingly finding more use within mathematics and computer science.

instance, today is sunny or it isn't. There exists such a thing as the Sun or doesn't. Mathematics of this kind of knowledge was first worked out by George Boole back in 1847 out of which Boolean algebra was born. The formal logic that works solely with this kind of knowledge is called propositional logic, also known as Boolean logic for obvious reasons. In propositional logic you have a set of symbols with Boolean values and a number of logical connectives (such as Boolean operators; AND, OR, NOT). Using the two you can form sentences that are also assigned Boolean values, either true or false. If you combine correct sentences to form new sentences, the new ones will be correct too if you follow the rules of propositional logic. And that's where you can see how it can lend itself to building AI systems.

Propositional logic assumes the world is made of propositions that are either true or false, and that's it. This lack of sophistication is a problem because for every instance of everything you must have a distinct symbol and for every instance of everything you may want to say you have to write a new sentence. For instance, suppose we want to express the notion that "everybody is at some place at any time" as a general rule or valid pattern. In propositional logic, I would have to spell that out for every single person-time-location combination. Besides the syntax and grammar of your logic, you would practically need some naming convention too for all the many symbols for every instance. Here's an example sentence "PERSON_A & TIME_t1_ofA & LOCATION_l1_ofA" and on it goes for all true combinations. It is just a bad way to go, although there are still systems in production that work with 10s of millions of sentences like this, doing just fine.

It is not possible to express any general aspect of the reality we experience in a framework solely made of propositions. And as we'll discuss later, this inefficiency is not limited to Boolean-logic-based systems. Many current state-of-the-art and not logic-based AI systems share this ugliness too including current deep learning systems!

To express any general rule, we need a formal logic that is much more expressive than propositional logic. This more expressive and general logic is called *predicate* logic developed independently by philosophers Gottlob Frege and Charles Sanders Peirce. Their specific formulation was in fact only a special case of predicate logic called, *second-order* logic. What has lent itself more to AI, is what's known as *first-order* logic that is the most elementary form of predicate logic. Even though the foundations were already set, the development of first-order logic should be attributed to David Hilbert, (whom we met in chapter 3) and Wilhelm Ackermann. Their 1928 book which translates to "On elementary mathematical logic", formulated first-order logic in the way that's been understood and used ever since.

We can indeed express our example of "everybody is at some place at any time", also in one sentence in first-order logic. It will read something like: for every person p, and every time t, there exists a location l, such that; Person(p), AND, Time t, AND, Location l, are true.[70] As you see, with first-order logic we can pack many propositions into one, saying something is true for all objects in the world even if we don't yet know about the existence of all the objects. It's a different way of looking at things, going beyond instance-based propositions and just making statements about relations/interactions between objects. That sounds a lot like how we experience the world. Indeed anything we could formally translate from English to first-order logic would take up roughly the same size, in terms of the number of sentences.

What about second-order logic? You notice that the example above (of first-order logic) was stated only about persons whereas, we know, the fact being represented is true about any object. So in second-order logic, we can treat relations and properties of objects as objects too and make general statements about them. For instance, we can say any object regardless of what property it has (its shape, color, etc., or whether it's human or not), is at some place and some time. We write (in second-order language) something that reads: for any properties P and any objects x, there exist a time t and location l such that, If P of x is true, then "Time t" AND "Location l for x", are true.[71] We haven't yet found a way to utilize second-order logic in AI. Higher-order logics are even more powerful such that our only concrete examples of them are in esoteric abstract theories of mathematics such as in category theory.[72]

There is some essential importance to this order-based categorization of predicate logic, but that's certainly not everything in logic. As we just saw, expressiveness and convenience matter in a language of logic. For that reason we can invent special-purpose logics too in order to express some special class/domain of reasoning patterns easier and more native in that domain. For instance, if you think a "time axis" is a priori to everything else and you want time-ordering and causality to be respected above all in your reasoning, you need a logic that is more amenable to temporal patterns. That is exactly what *temporal logic* aims to achieve. Other special-purpose logics introduce other operators that make things more convenient just like temporal logic makes time-ordering a first-class citizen in temporal reasoning.

---

[70] $\forall p,t \; \exists l : \text{Person}(p) \; \& \; \text{Time}(t) \; \& \; \text{Location}(l,p)$.

[71] $\forall P \; \forall x \; \exists t,l : P(x) \rightarrow \text{Time}(t) \; \& \; \text{Location}(l,x)$

[72] For instance, formalization of closure under union operations axiom in Topology requires third-order logic.

Putting convenience and nativeness of logic aside, all types of logic used in AI, be it Propositional, first-order, or any special-purpose variants, share a fundamental feature which is THE reason why they are used in AI. That is, as we said, they are formal and fully general with respect to the subject matter. In short, formal logic is humanity's only attempt to objectively define and guarantee the "correctness" of reasoning. It's hard to imagine anything that could not be expressed with predicate logic since one can always go to a higher-order logic to accommodate higher expressiveness. That's not a proof nor are we close to needing a proof (at least, when it comes to AI). Predicate logic is already capable of expressing patterns way beyond our typical imagination. The problem with the current known forms of logic is not a lack of expressiveness.

As we said, logic started with Aristotle to formalize all the spoken. Even if we could expose all those patterns formally, there is still a lot that we never say. Some of which we are not consciously aware of and some of which are not even at the level of our intuition. The problem with logic is all the things we don't say, everything that we don't know we should express, for a completely formal reasoning that would be useful for AI. Thanks to our fundamental bias, we take many things for granted since we all share very similar brains that help us do things the same way, talk the same way about things, and have the same distorted view of what's out there.

One thing we have no direct access to in our brains is the concept of probability, it's not a first-class citizen for us humans. Most of the irrationality we spoke of in the previous section, that psychologists and economists attribute to humans, can be traced back to this fact. Earlier we said logic in AI works with definite information. By definite, we meant that every symbol of statement could be assigned a true or false value, unlike in reasoning with uncertain information that is typically considered to be the domain of probabilistic reasoning. Probabilistic reasoning is distinct from logic, and an act of inference can either be called logical or probabilistic.

There are different kinds of uncertainties, however. We are not equally unfamiliar with all. What we are natively unfamiliar with is mainly just about the existence uncertainty, the probability of occurrence or existence of something that we are at least clear about its identity. If something exists for sure and we are certain of its identity, we can still be uncertain about our description of it. Our description could be inaccurate and fuzzy. That is something we understand directly from our senses, we can tell whether something is there or not with certainty but it may look or sound fuzzy.

Fuzzy logic takes on representing reasoning patterns with this type of uncertainty. In Fuzzy logic, we can work with *fuzzy sets* where elements of the set can belong to it with some uncertainty.[73]

For instance, suppose we see a picture and we know it's a picture of a cat with a 100% probability, but we may not know the exact boundary of its body. We could quantify that uncertainty over boundary pixels with fuzzy values (not just 0 or 1 but also in between) assigned to every pixel indicating to what degree each pixel belongs to the cat body and not the cat body (two fuzzy sets).

There are other uncertainty measures too, generally referred to as *monotone* measures, but these two kinds of uncertainties, probability theory and fuzzy logic, already cover almost all uncertain natural phenomena. Yet, there remains some philosophical gaps. We could call these gaps "expressional" uncertainties; uncertainty in the degree of truth of a judgement. For instance to distinguish between possibility and necessity. Such expressions as "it is necessary that..." or "it is possible that..." are called modals. They are used as quantifiers in what is the most known form of *Modal* logic. It is an example of a much more appropriate uncertain logic for many forms of philosophical reasoning than, say, fuzzy logic or probabilistic inference.

Lastly, setting aside the problem of things we don't say or don't know exactly how to say, there is an obvious problem with logical reasoning, namely what if our premises, what we do say, are wrong to begin with? There are attempts to combat this obvious failure mode too, at least when it comes to AI. *Non-monotonic logic* is an example, in which logical deductions are allowed to be withdrawn in case there comes any reason or evidence to doubt any premise.

Pre-Modern to Modern AI

Covering the birth of AI in chapter 3, we mentioned that the ability to solve problems in mathematics and performing logical reasoning were mistakenly deemed to be the core of exhibiting intelligence, that's how AI got started, instantiated in a number of programs. Now that we have enough background in what logical reasoning is, we are better positioned to continue our dense review of history to arrive at the birth of modern AI.

We mentioned Alan Newell and Herbert Simon's general-problem-solver and the logic-theorist. There were other programs too, similar in style, that we did not mention, such as geometry-

---

[73] This uncertainty is quantified by a membership function that says to what degree (between 0 and 1) an element belongs to the set.

theorem-prover at IBM, or John McCarthy's advice-taker which marked the birth of Lisp[74] (that would come to dominate the world of AI for almost three decades).

One notable attempt, which came by even earlier, was to solve the problem of what series of moves would result in a win in the game of checkers. Given limited computer memory, especially in the 1950s, evaluating all strategies for playing is not practical. Arthur Samuel wrote a program at IBM to evaluate the goodness of a possible move without going through the whole tree of possibilities. This evaluation could become better the more variations it saw, i.e. the more the program played. He called his algorithm rote learning. He also used the phrase "machine learning" in 1952 to refer to what was happening. That was the first usage of the phrase.

His winning checker's program was and still is quite remarkable given the extremely limited, compared to today's standards, resources of memory and compute which he had to cope with. However, what was ignored was that these were all examples of cases where 1) the problem to be solved could be written down mathematically and precisely, 2) the operating environment is fully known with deterministic rules about how it operates, and 3) deterministic outcomes for every possible action. The challenge to overcome in these AI programs was exactly how to represent the problem (how to make the computer "speak" the language of that problem), and how to find the solution in a huge space of possibilities (where each configuration contains the full description of a state of the environment and the state of the agent).

A good example is the game of chess. Suppose you've started to play. Right now, the position of everything on the board, that is, and whose turn it is to play, all together form the current *state* of the game. All this state information can be represented by a node at the root of a tree. There are many possibilities to come next, based on the moves you and your opponent make. Different possibilities (resulting in different states) are represented by nodes on different branches of the tree. This is called the "game-tree", where the leaves of the tree (the final nodes) are states where the checkmate conditions are met. So your task is to find a path on this tree that goes from the root to a winning leaf.

This is true not just about chess but any game with known static rules. The game-tree could be huge. The bigger it is the harder it is considered "solving/winning" it. The size of a game can be measured by the average branching factor over a statistically large set of games and moves — the average number of true possibilities per move. For checkers, this factor is about 3, for chess about

---

[74] The language where self-reference is explicitly designed for, in a way that could allow a machine to "think about itself". For that reason, many people are still in love with it.

35, and for GO it is about 250. That's why chess wasn't solved in early AI and why GO is considered a lot harder than chess.

After almost a decade came expectations of doing something practical and commercially useful. The period 1965 to 1975 marks the era of such attempts. The late 60s involved lots of conception and R&D work. By the early 70s, there were AIs in production, roughly for two different types of applications: 1) predictions using highly specialized knowledge such as in science and medicine, and 2) planning and operations.

Expert systems came to address the first category, where the AI developers would interview experts and go back with their notes to work on representing that knowledge in a knowledge base — think of it as a bunch of logical statements (starting with symbols and definitions) which you can operate on with the rules of logic. Using their knowledge-base and some premises as input they could deduce new facts. Consider MYCIN, the first successful expert system, developed to diagnose blood infections and to suggest treatments for them. Although it was never used quite in that way in the real world, it sparked fruitful development of similar systems in other domains outside of medicine.

In the second category, applications in planning and operations, there were prominent examples too. For instance, many planning problems in logistics can be formulated as a constraint satisfaction problem. Such AIs would systematically search for viable solutions that satisfy the constraints, and prescribe the solution as an actionable plan. Logistics planning systems in production were basically a set of propositional logic systems. They are still in use to this date, except that the modern versions maintain millions of symbols and can process 10 times that, roughly 10 million propositional sentences. Similar systems are in place inside even the high-tech industry e.g. for software release and software security checking.

The next decade, roughly from 1975 to 1985, these technologies went commercial circulating through the industry as the coolest, latest, and greatest. With wider use, the problem with purely logical systems which we covered in the previous section started to surface itself. The first issue was, per our discussion in the previous section, all the things that we don't specify, or forget to specify because they are just common sense! For instance, MYCIN had no idea that there are concepts such as doctors, patients, hospitals, relationships, and restrictions therein. These kinds of observations in celebrated programs like MYCIN, eventually led to a movement in AI to try to build machines with common sense. The most notable and monumental effort was the Cyc project, started in 1984, and surprisingly some people are still working on it.

The second issue, that was widely noticed, was that the knowledge in these systems were represented by absolute statements with no flexible way to represent uncertain information and to calculate probabilities. Of course, developers of these systems all knew about probability theory. What they didn't know yet, was efficient methods to represent probabilistic knowledge, and to perform probabilistic inference.

Academic research on those fronts was ongoing in parallel. In 1988, Judea Pearl wrote a convincing book for AI researchers to adopt a probabilistic framework for AI agents, namely, the framework of Bayesian networks. Therein, he also introduced some efficient inference algorithms on Bayesian networks. However, the adoption of probabilistic agents came only in the 90s which was too late to save the AI spring from the growing pessimism. The term AI winter was first brought up after the public debates in 1984 on AI. Soon there were severe and abrupt setbacks in funding in both industry and academia. That had a major hand in the collapse of the Lisp-machine markets starting in 1985.

In the absence of good probabilistic frameworks, practitioners in the 80s who were building the rule-based AI systems (of which expert systems are a special case), went on to represent uncertainty with new rules or meta-rules (rules about the rules) that would put varying degrees of confidence on the certainty of a rule or logical statement. These are not probabilities but some ad-hoc representation of uncertainty by numbers that may bear some similarities to probability values. An Infamous attempt in this direction was the use of an uncertainty framework known as *"certainty factors"* in MYCIN.

Such ad-hoc methods for representing and quantifying uncertain information are still widespread in the tech industry, for the simple reason that they are a lot less work and less resource-hungry than any rigorous probabilistic inference. For many applications, this may turn out to be just fine, and much better than no degree of uncertainty representation. However, for the promised quality of AI systems, such attempts were far from sufficient to save the day.

Starting in the 90s, people started to realize that agents MUST be probabilistic in nature. That marked the birth of modern AI. Previous (symbolic rule-based and logic-based) AI systems were categorized as GOFAI, good-old-fashioned-AI. The association of AI with logic turned out to be better for the field of logic than the field of AI. Logic started to benefit a lot from the work and attention of computer scientists, not just the traditional crowd of philosophers and mathematicians working in logic.

A great example is the development of the language prolog, which is still widely used in the industry. Even the notoriously secretive IBM Watson uses prolog programs in its stack or at least it used to. The development of fuzzy logic and fuzzy-logic-based AI systems is another example — also still in abundant use, especially in Japanese manufacturing, from washing machines to rice cookers.

GOFAI was not scalable in the sense that it required too much human effort. It requires too much human expertise to be put into the system in the form of rules. That is the so-called "knowledge acquisition bottleneck", which makes building a machine by putting our common sense knowledge directly into it too hard. Even worse, some applications may demand specifying some tacit knowledge that we don't even know how to specify, or how to specify sufficiently for the system to work well.

This last part was exactly what led to the adoption of *machine learning* in the 90s — building systems that learned directly from examples. Machine learning (ML) is the topic of our next chapter, but while we are on the topic of non-scalability due to the amount of human effort required, let us mention that ML systems were not totally scalable either. One still has to gather a lot of data in the form of useful examples, clean them up, and hand-craft the features (that would represent the examples in the algorithm) in a process called feature-engineering. Those are amongst design choices for the developer, instead of the AI itself.

Although deep-learning promises to get rid of that need for feature-engineering, it is not yet applicable to many practical problems and demands of the industry. Therefore, handcrafting knowledge in AI (and analytics) is still a problem preventing systems from being scalable and generally applicable.

Chapter 5

AI as Machine Learning

Before getting into what "learning" refers to exactly, let's continue with our semi-chronological storyline. With the rise of the internet industry in the 90s, new kinds of data were being generated by the digitization of many forms of content for online users, and by the activities of those users. With that, also came new data-based technological opportunities beyond what were once only in the domain and hands of database technologists.

These opportunities were seized by those who wanted to do a lot more with that data. Machine learning (ML) and applied machine learning in the form of "data mining", started to gain popularity because of that.[75] The scalability issues with pre-modern AI systems (discussed earlier), led to almost all previously state-of-the-art systems being replaced by ML-based systems over the past two decades. The appeal is obvious, instead of having the AI designer anticipate all possible contingencies and directly include some knowledge for the system to handle each possibility, one just provides a lot of examples that implicitly contain that knowledge. That way, we can talk to the system through examples only, rather than by specifying the human knowledge implicit in those examples.

Still, it doesn't mean that every possibility will be handled in a given ML system. There could be many important so-called edge cases left over, where the learning system (with all its design and data), either doesn't anticipate or fails to handle. This can be due to known-unknowns or unknown-

---

[75] To this day there are two main kinds of technologists in the data-based technology industry, those who evolved from the world of databases, storage and high performance computing, and those who come from the statistics, analytics, and math background. Though, there are many practitioners who are effectively well-versed in both.

unknown that show up after deployment of the system, not caught by the pre-designed tests during the development process. This is indeed an incredibly challenging part of what an AI engineer must attend to in the world of modern AI. However, it is still much less and almost not comparable to the set of possibilities that say a GOFAI engineer had to contemplate and code for.

For instance, the most naive way to handle the edge cases of an ML system would be to pass it through a filter that is made of a set of rigid rules in the form of: *"IF* the conditions of edge case X are satisfied, *THEN* replace results by result-set-X", so on and so forth. Compare this to a naive rule-based GOFAI system where this is not just about handling edge cases but all possible cases! Worse yet, in some domains, humans are incapable of representing the necessary knowledge, with sufficient accuracy, and also explicitly in the form of logical statements — in processing audio signals of speech for instance. This explains how the discipline of machine learning came to dominate AI quickly.

In the old world, all the responsibilities were on the AI designer to basically figure out everything without any prior theories, or principles. The only test was if the engineered system worked! The job was to simply "do it right". The situation for ML was a bit better. There are at least some theorems that make general statements about when your design (hypothesis space) is able to learn from the examples, in terms of how sophisticated your system is and how many examples you need, to be sufficiently close to the right answers with sufficient confidence. Such statements are all part of statistical learning theory and its extensions, which we'll review shortly. Before getting there, we have to talk about what the heck do we mean by learning?

It can be confusing because it's one word representing different concepts in different domains/contexts. People from various backgrounds, say, in computer science, cognitive science, biology, philosophy, and folk wisdom, all bear in mind different notions (though somewhat related) of learning. Nowadays ML is widely understood, but just a decade ago, I used to witness mixed-background groups totally talking past each other on the topic of learning and yet magically ending up with mutual head nods! Interestingly, on the flip side, when different disciplines have espoused the same notion of something, they tend to give it totally different names.

For instance, the discipline of signal processing and computer science both carry the same notion of learning, except in signal processing it's called adaptation, and the learning system is thought of as an "adaptive filter" which you can interpret as broadly as you can a learning system. Statistics too, carries the same notion although with a slightly different view on the goals and there it's typically called "parameter estimation" and almost never called learning.

Computer Scientists wanted a slightly different name too, for the similar reason that the term "learning" already existed in psychology. What did they do? They called it *Machine* learning! Arthur Samuel came up with the phrase in 1952, but the first push to popularize it came only in 1959.

The shared notion of learning in computer science, statistics, signal processing, and other nearby engineering disciplines concerned with the design of autonomous and robust systems, was cast in a precise mathematical formulation independent of machines or computers, and is known as the learning problem. The learning problem was known decades before ML became popular. Unlike this well-defined learning problem, the term machine learning is a quite loose afterthought, similar to the label AI.

Being a loose term, we did give a loose definition in chapter 1, which was a watered-down version of Tom Mitchel's definition,[76] who wrote the first widely popular and comprehensive book on machine learning called "Machine Learning"! Here's another definition from the most popular ML course in the world offered by Stanford[77]: "ML is the science of getting computers to act without being explicitly programmed"! Of course, that is an ambiguous statement as to what counts as an act or explicit programming, but it simply doesn't matter how we define it because its purpose is only to give a sense of the territory it intends to mark. The bottom line is, if your program uses data to improve its performance in any way, you can call it ML, and if there is any ML in your stack, you can call it AI. Well, at least for the time being until we get more serious about defining intelligence.

Statistical Learning Problem

The learning problem we're considering has nothing to do with the one in psychology, it is about the theory for all the engineering fields that are concerned with building (semi-)autonomous and robust systems, be it in machine learning, signal processing, control, etc.

Curve-Fitting vs. Learning From Examples

---

[76] "A computer program is said to learn from experience E with respect to some class of tasks T and performance measure P, if its performance at tasks in T, as measured by P, improves with experience E." Tom Mitchel, Machine Learning 1996.

[77] Andrew NG's course at mlclass.stanford.edu

This definition of ML we just gave, i.e. using examples (called experience) to increase performance is a bit problematic. For one thing, just a (look-up) table of examples can increase its performance given extra examples, stored as new rows in the table, in the sense that it can answer more questions, those questions that can be entirely answered by retrieving the new examples. This table alone can say nothing about an example that is not stored in it. So it seems like we should be requiring a learner, something about an ability to generalize beyond the examples it has seen/stored.

Before we get to the "learning problem" formally, let's discuss the gist of it. The learning problem is a general formulation of the problem of "fitting a curve" to a number of points (your data points).[78] This is not a metaphor, it is exactly what we do in Machine learning. So if it is just "fitting a curve", why do we call it learning?

"Fitting a curve" was never recognized as a fundamental problem it actually poses. Perhaps that's because of the traditional contexts of its application. First, traditional use was almost always where we already had some theory about the "curve" — the data would come from experiments that were guided by a theory about the data. The a priori theory contains some properties the curve must satisfy, which would narrow down the family of curves it should belong to, and different members of the family would correspond to different parameters in the theory. Second, in traditional curve-fitting, finding the right member of the family is cast in a rather simple optimization problem. Third, it is typically in contexts with very low number of dimensions compared to that of typical ML problems.

These are the main reasons why we never took the problem of "fitting a curve" seriously enough. To most people's ears, fitting a curve is just an optimization (the second reason above). But one must note that choosing to formulate fitting as an optimization, is not an a priori must, though that's what has always been mistakenly assumed. Similarly, some people may feel the same about ML, that it is just some optimization problem.

Whether it is curve-fitting or learning from examples, the challenge is really about *generalization*, exactly in the sense of inductive reasoning, going from a finite set of data points to

---

[78] This is something that most likely any high school graduate has experience with. In two dimensions you have a curve, in 3 you have a curved surface, and so on. In general, you have a *hypersurface* (a dimension-independent name), an n-dimensional surface inside an n+1-dimensional space. For instance, You gather salaries of some employees of a company, and their rank. You have a number of points on a plane and you can fit a curve to them to be able to have an estimate of salaries of other employees if you knew their rank. Now if you add the variables "years at the company" and "industry-sector of the company", you would have to fit a 3-dimensional hypersurface to your points.

say something correct about a data point you could observe but haven't yet. That deserves to be called a problem, i.e. the learning problem. Because it is not one that can be entirely and generally solved. It's tainted by the fundamental (Hume's) problem of induction, or more precisely, the artificial problem of induction, following our discussion in rethinking rational agency. So instead of solving it, we make it not be a practical problem with high statistical confidence, using optimization. That is the gist of the statistical learning problem.

How is learning different from curve-fitting? It isn't, though, given the three above-mentioned restrictions in the traditional contexts of application, we can say *learning is generalized curve-fitting*. In the problem of learning, we don't know the correct family of curves with certainty.[79] The optimization problems we form for the purpose of learning are not restricted by any means. The number of dimensions can be very high and the geometry of the space can be endlessly complex.

For some experts, this is not enough of a difference. Judea Pearl whom we met in chapter 4, calls current mainstream ML practices, the exercise of curve-fitting because the models don't know any "why", that is, they don't have any notion of causality, which for him is a criterion to go beyond just "fitting a curve".[80]

Having said all that, in the general problem of learning it is possible to actually bypass fitting a curve and try to only find a way to say something correct about a target question. Explicit bypassing of fitting a curve (induction) is sometimes referred to as *transductive* learning. Transduction is something between induction and deduction. You go from specific statements to other specific statements (unlike induction) that are not logically entailed by the initial statements (unlike deduction).

Types of Learning

There can be many different settings for a learning problem. Different settings may practically require distinct approaches. So it should be obvious that there are many different types of learning systems within ML, and frankly, there are too many names associated with different settings, which in turn, may be signaling nothing but the infancy of the field. We will refrain from using these names

---

[79] We may have a prior belief that quantifies our variable confidence in different families, which we can capture in our learning procedure — this is done most rigorously in Bayesian learning.

[80] See "Book of Why" by Judea Pearl.

as much as possible, instead we mention some of the causes behind the multiplicity of names. Here's a list of the most outstanding aspects of learning that's used for categorization of learning systems:

1. From what you learn? what feedback is available to learn from? (supervised vs. unsupervised vs. reinforcement learning)

2. How do you learn? How are you representing the learning problem? What's your approach to learning? (different approaches and camps)

3. When do you learn? When are you going to make up your mind how to use the example/experience for inference/decision/action relative to when you receive the example? (lazy learning vs. eager learning)

4. Can you ask questions when you are in your learning session? (active vs. passive learning)

5. Do you have time to get all your examples before answering questions or do you have to be answering questions meanwhile and learning from them? (online learning vs. offline from a batch of examples)

6. How much learning is actually happening for the target task? What component of the AI agent is using learning as opposed to, say, getting by with pre-set rules?

7. What are you leaning for? For what kind of task and scope in performance? Are you learning for a single prediction (e.g. learning only to *discriminate* between cases) or are you preparing for other tasks related to the same data distribution (e.g. *generating* new cases and examples, or transferring some of what you've learned toward performing yet-unknown tasks)?[81]

8. Where do your lessons come from? Are your examples coming from an environment that is set out to fool you, or a caring and helpful teacher? Are the examples you see independent enough for your statistical learning?

9. The type or representation of examples? Format and geometry of examples, e.g. graphs, sequences, grids, etc.?

10. Do you build your learning upon any prior knowledge or not?

---

[81] In representation learning, depending on what kind of downstream tasks you would want to be able to transfer your model to, you'd need to perform different types of learning e.g. by modifying learning objectives so that you can learn better *posterior* distributions, not necessarily optimal for a current task at hand, considering trade-offs towards being good at different tasks.

11. The level and type of interpretability of the learning algorithm.[82]

Indeed an enormous variety of learning systems have been studied in the literature. So the above is not an exhaustive nor an ordered list. Plus these aspects are not on par with each other in terms of their ability to separate learning systems. Many of them are typically orthogonal aspects and one can use them in conjunction to classify a learning system, such as with "from what you learn" and "how you learn". There are also cases that different dimensions do overlap and correlate, such as with "How you learn" and "When you learn". Some of these aspects are worthy of categorization, others can be quite superficial at times. We will open up only the first aspect in our list above, i.e. namely categorization based on feedback since it is the most popular narrative in ML. That is the division of learning problems into *supervised* learning, *unsupervised* learning, and *reinforcement* learning.

Let's first get a bit more clear on what constitutes "feedback" and what constitutes an "example" to learn from? An example (which is also referred to as an experience of a learning algorithm) is basically a unit of data used for learning. And the set of all these units of data is the data-set. What is often assumed in most ML algorithms is that each is just a point in a very high-dimensional space (or a sequence of points in a slightly lower but still very high-dimensional space), represented by a vector whose elements are the coordinates/attributes of the data point along each dimension. We shall assume the same. However, as a side note, we should stay mindful of the fact that this is not an innocent assumption, that is, different dimensions may not be independent of each other to the extent that the overall space may not be a well-defined (metric) space (one that there is a meaningful distance between different points in it).

Feedback refers to the labels or values associated with these points or sequences of points. So it is part of the example, except it's carrying information about the output variables or target of the learning system.

Supervised Learning

The examples are seen as an input-output pair. Input being that vector (or vectorizable set) we just talked about and output is the label-set (typically just one symbol) on those vectors. Generally, if the

---

[82] For instance, (hard) decision trees, soft decision trees (hierarchical mixture of experts) and multiple trees (mixture of hierarchical experts), each offer a different type of interpretability. Nowadays, it's fashionable to speak of explainability instead, under the flag "Explainable AI".

output is a finite label set, we call it a *classification* problem, and if it could be represented by an infinite set, the problem is called a *regression*. The simplest case for an output is a binary variable, in which case the learning problem is a binary classification.

In all cases, one must find a function that maps the input space to the output label-set or infer probability distributions over the input space conditioned on a given output. Supervised learning means finding this function or inferring these distributions regardless of whether you find them explicitly or not, as long as you have access to them to generate a value (prediction) or infer a probability value for the output given any input from the input space (probabilistic prediction with conditional distributions). That is, you can predict or infer the likely label-set for any input, especially unseen inputs using only the examples you have seen so far.

Supervised learning is more understood and developed than any other ML problem. It also amounts to almost all industrial successes of applied ML. All the fuss in the media over the last couple of years about AI is about prediction using supervised learning. This may be partly justified because functions are everywhere — see Appendix B. Not only there are many outstanding problems that are begging to be solved by function approximation (low-hanging fruits), many other problems can too be cast as supervised learning problems. The challenge is of course having the data, the example pairs.

Producing, gathering, or finding and buying raw data is far from trivial nor the final step. "Cleaning" the data and labeling it properly isn't simple either. Labeling is the part that often gets outsourced. People get paid to label the data, sometimes working in data-set production farms! Still, further cleaning is needed by data scientists who build the learning systems. Yet often they cannot do much about improper labeling (noise in the ground-truth), as currently, popular ML algorithms are not that robust to such noise. Thus, perhaps the phrase "data is the new oil" bears a lot more truth to it currently than "AI being the new electricity".

Reinforcement Learning

The reinforcement learning (RL) problem is different from supervised learning only in that:

1. The supervision/feedback is not exactly the labels but only a clue about the labels.
2. Unlike supervised learning, the learner's choices influence what examples will be seen by it.

Otherwise, just as in supervised learning, in RL one must also find a function that maps inputs to outputs using some examples. The inputs are the situations (the states) the agent finds itself in and the output is the action to take. The examples don't include what the correct output labels (the actions) are, but it does include occasional numerical signals known as a reward value (or a cost value) coming from the environment due to the actions taken by the learner. Each action may yield a different reward that gives the learner a clue as to what actions may be better in service of maximizing the total reward. The goal of the full reinforcement learning problem is to maximize the cumulative reward from a sequence of actions.[83] If there is no terminal state, the task may continue forever (*continuing* tasks as opposed to *episodic* tasks), the length of the sequence of actions has no natural bound, such as in financial investing. In this case, in order to have a well-defined learning problem, one must enforce this cumulative reward to remain a finite object and that is often done by discounting future rewards.

The second difference with supervised learning is also fundamentally distinctive, in that you have a choice in what examples you see. Examples are generated sequentially based on what the agent has tried so far and what it chooses to do at any given time based on how the reward pattern has been shifting. Different choices yield different examples for the agent to learn from. A fundamental trade-off to balance out, in making these choices, is between choosing from likely good actions you have tried so far (exploitation), or risk it by trying something new in hopes of something potentially more rewarding (exploration). Such a trade-off is non-existent in settings of supervised or unsupervised learning (which we review next) since the learner has no influence on which examples it sees.[84]

What we have described so far is a very general problem setting. We have made no statement about how one should solve or approach this class of problems. The name "reinforcement" can be misleading, as it is inspired by an approach to the solution, not the problem. The approach is that of trial and error, while trying to positively "reinforce" the rewarding actions (responsible for the reward) and negatively reinforce the costly ones. How to do this properly, can get endlessly complex. For instance, in many practical contexts, we may not actually want to try anything online before learning about the problem or its components in an offline manner first.

---

[83] In more restricted versions of the problem such as in bandit problems or associative search, the rewards and actions are only local with momentary effect, i.e. can be thought of as 1-action-long episodes.

[84] Though one can introduce a trade-off of this kind in the context of active learning.

That is a necessity, since online trial and error may result in actions that are too costly — it may result in the death of the learner. In this sense, there may not be anything there to "reinforce" (in the traditional sense), yet you must solve the learning problem. Therefore, not only the agent doesn't get to forget any seen example, but it has to properly learn a more accurate model of its environment and itself (typically referred to as model-based RL), in order not to try many things and this barely has anything to do with the process of "reinforcement". It is perhaps more interesting to view a reinforcement problem as a problem of "on-line" planning.

Progress in RL techniques has caught a lot of attention in the media as they involve a notion of action that is easier for people to anthropomorphize. Currently, when it comes to systems in production on real-world problems, these advances RL are an insignificant fraction relative to plain supervised learning systems. One reason is that they need a lot more examples to learn since the feedback is a lot weaker than in supervised learning causing the learning to be a lot slower.

In problems where one can simulate the environment with sufficient accuracy, examples can be generated endlessly and an RL agent could be developed well. That's obviously true in the case of computer games and that's where the main show of progress has been. A progress that's been based on 1) the use of deep learning techniques, and 2) larger and cheaper compute power. Again, for the most part, the progress in RL research hasn't been graduating over to industry, something that'll most likely change in this decade.

Unsupervised Learning

Learning without any external feedback (be it labels, associated target values for the examples, or any information from the environment that the learning is on the right track) is a much broader and much less understood category of problems. This is the category of unsupervised learning problems. However, it can still be understood as function approximations or inferring probability densities which are staged for human experts to easily label different parts of them according to what's useful for the target downstream tasks. Let's unpack that a bit.

In what sense is it still function approximation? Without labels or feedback one can formulate the problem as inferring, not the conditional probability distribution as in supervised learning (to directly *discriminate* between labels), but the joint probability distribution over the input space, the distribution where the examples are coming from. Unsupervised learning at its core, is about inferring something about this probability density. In particular, inferring special structures of this density function.

One typically assumes that there are some *latent* variables that are behind such structure—when the assumption is explicit the model is called a latent variable model. For instance, the structure could simply be the result of an overlap of several joint distributions that are (conditionally-) independent given the latent variables. Therefore, we can introduce auxiliary labels for the inputs to be jointly learned with the inputs. The attempt is to capture relevant latent variables by these auxiliary labels which, if not the cause of the structure of density, could at least provide a way to explain it, and also a way to generate new examples compatible with the overall density of the data (known as generative modeling).

Learning about the joint probability density of data can be cast as a function approximation problem. If we can learn about the probability density of the data and/or its structure, there is a lot more that we can do. We can learn better representations of our data that are more refined and distilled (e.g. with lowering dimensionality, pronouncing the relevant parts, and suppressing less-relevant parts) for porting it over to other tasks. We can use it to understand the noise in our problem better, make it easier to discover patterns, and discover anomalies. We can use it to group examples and data points into different clusters and learn about the distinct properties of each — that's called a clustering problem.

Having said that, to go from an approximation of the joint distribution of our unlabeled data to these well-known applications of unsupervised learning, like anomaly detection, clustering, or detecting a trend change in time-indexed data (time-series data), is not trivial. That is, any jump to any label brings about a subjective effort by a human (expert in that domain). In supervised learning, the label comes a priori making it easier to forget about the noise in that label, whereas in unsupervised learning, the labeling comes after learning (at application time), and it's harder to ignore its downstream consequences. To make this point clear, let's take the specific case of clustering.

The most relatable form of clustering is categorization, which constitutes a huge chunk of what taxonomists (or scientists[85] for that matter) do. Suppose we have the true probability density of the data (ignore that finding this density explicitly is often an intractable problem), and that there are

---

[85] You can think of what one does in science and knowledge production in general as either an act of classification (put new cases and phenomena in one of existing clusters) or an act of clustering (categorize the new case as a new cluster since it's qualitatively different than all you've seen before and give it a new name). Although, technically speaking, this would be a form of non-parametric clustering, with the added requirement for an ability to merge clusters when needed.

some valleys in this distribution. Placing contours in these valleys gives us a bunch of regions trapped between the contours. We can give different names to each region and call them different clusters. We can do a pretty good job in this act of clustering if we know the target application of this grouping, or if we know the latent structure of the generation process of the data (where the examples/samples come from). But what if we didn't, which we generally don't? Isn't there always some element of taste involved here? Should we expect there to be a perfect clustering algorithm? Intuitively no, simply because of the presence of subjectiveness.

This intuition is formally captured in a theorem proven by Kleinberg (2003) called the "impossibility theorem".[86] Although one can ease some of the criteria in this theorem to achieve good clusterings, the point is that there is nothing fundamental (human-independent) about the act of clustering beyond potentially capturing some true latent causal variables. Similarly, anomaly detection can be seen as a special kind of clustering where you have one very large cluster (everything normal/non-anomalous) and one or more tiny clusters corresponding to different types of anomalies.[87]

The progress in the general problem of unsupervised learning remains mostly academic, though promising. Most of the supervised learning usage has been in service of some overall supervised learning. That is, unsupervised components either used for preprocessing or embedded inside a supervised learning setup to reduce the number of labeled examples needed, as well as to increase the quality of learning. Just like the problem is a general one, we seek also a general solution to it, rather than a specific method for a specific instance of the problem, say, by leveraging domain knowledge. Perhaps truly unsupervised learning has more to do with intelligence than the previously mentioned categories of learning, which is why it is the dream of most researchers to be able to make significant progress in it.

The Learning Problem Formally

Earlier, in the discussion of supervised and unsupervised learning, it may have seemed like we were unnecessarily distinguishing between approximating a function and inferring a probability distribution. We know that inferring a probability distribution is a special case of function

---

[86] "An Impossibility Theorem for Clustering" Jon Kleinberg, NIPS 2003.

[87] Keep in mind that anomaly detection can also be seen as a combination of supervised and unsupervised learning if you have some labels supervising you directly on what some of the anomalous examples are.

approximation [see appendix B for a review of functions]. Probabilities and probability densities are special functions with properties that are necessary to give theoretical meaning to the act of prediction. There are of course many non-probabilistic approaches to prediction in ML. Those too, find their meaning in relation to some probability distribution, as in they are supposed to be "close" to the peak of some probability distribution.

The assumption in the statistical learning problem is that there exists a true probability distribution where the examples are coming from. With that, we can unify all different types of learning under a general setting of the learning problem. Given that all different learning types involve some function approximation (e.g. estimating various conditional or joint probabilities), we can abstract away the (superficial) differences between the types, and work on the general aspects of the learning problem independent of the learning type (and its setting).

To that end, assume there is a probability density of examples; name the examples z and the probability of z as $p(z)$. The form of z is different for different types of learning of course. The degree of goodness of the act of function approximation at every point is measured by a function called a loss function (also different for different learning types). Loss function takes a point in input space (some z) and your function approximation at that point, and is supposed to tell you how good you are doing there, the lower the loss there, the better your approximation there. How about the goodness of the overall approximation, the learning? The expectation value of this loss over the true data distribution (its average over an infinite number of z samples from $p(z)$), known as the statistical risk, will tell us that. It's called "risk" for the obvious reason that being wrong is being risky (bad things can happen when you're wrong). Minimization of this statistical risk is the formal statistical problem.

What we just described is a general setting for the learning problem, but it is only the *frequentist* version of the learning problem. There are of course *Bayesians* versions too.[88] We should note that not being Bayesian is problematic. The reason is that we never know the above-mentioned $p(z)$

---

[88] For those not familiar with statistical approaches, it suffices to know that there are two main approaches to statistics: frequentist and Bayesian statistics. They fundamentally see proper inference of probabilities differently. Although they would always eventually (after enough number of samples, often infinitely many) agree with one another, with finite number of samples they do have practical as well as philosophical differences. Bayesian don't like the fact that frequentists often talk in terms of things we cannot know (things we have no access to or that requires infinite number of samples, i.e. impractical) and frequestists don't like Bayesian methods, where one must declare some prior beliefs about the samples before having seen any (which can't be done perfectly, although practical and very effective in leveraging all we do know).

distribution, and thereby the statistical risk above is not computable. However, because we can approximate it, it turns out that one can still work out very useful theorems on the general aspects of the statistical learning problem sidestepping the differences in statistical approaches. So we'll work with the frequentist version, simply because it is a bit easier to describe, and conceptually sufficient.

There are the three main theoretical aspects of the learning problem that we must understand: *representation, optimization, and generalization*, which we'll discuss below.

Representation

Effectively anything from the examples to the loss function, including the model (the hypothesis set), each must have a mathematical representation, a representation in the machine that the machine knows how to work with. All these very many choices and how they may affect learning are studied in the representation theory of a machine learning algorithm.

The representation problem exists precisely because we don't know what function we are supposed to approximate, how complex it is, and so on. Complexity can be measured in many different ways, could be as simple as counting the number of parameters in the model, or much less naïve than that. Even with the right measures of complexity we still wouldn't know the actual function, and whether our hypothesis space (chosen by our representation) contains it. Assuming that it does is known as the *realizability* assumption, and relaxing this assumption is known as *agnostic* learning. It is possible to relax that assumption because we just need to figure out a good approximation to the target function. So the representation has to be just good enough for the desired approximation.

There is yet another assumption that we can relax. That is by not requiring that the learning algorithm chooses a hypothesis from our hypothesis class. We can start with our hypothesis class and then allow for the algorithm to find a better hypothesis from a bigger space that contains our original space. That is known as representation-independent learning, sometimes called *improper* learning.

Optimization

Finding a hypothesis within our space that minimizes the statistical risk is the task of the optimization algorithm. Once we have a representation for everything, we start from an initial hypothesis in our representative hypothesis space and let the optimization algorithm make that hypothesis better by applying a series of tweaks to it, and each time seeing how it improves the results.

The hypothesis space is always parameterized by a number variable, so when the learning algorithm outputs a good (or bad) result, it is unclear how much each variable's value was responsible for the result. That is called the credit assignment problem and is the main part of what the optimization algorithm has to solve. Think of looking at any part of the economy and trying to figure out which actions/decisions of which previous (including the current) governments led to its current state so that we can better support or reject future policies. There you have a credit assignment problem.

Since we have a mathematical representation of the hypothesis we can relate the change of the output to the change of any of the parameters of the hypothesis. The simplest form of this relation between these changes is captured by first-order derivatives, also known as gradients. That's what the most popular learning algorithm in neural networks uses to solve the credit assignment problem and keep tweaking the parameters until a good hypothesis is found.

Understanding what kind of optimization problem we are dealing with, how to solve it best, and how to verify that we have solved it satisfactorily, is what the optimization theory of a learning algorithm studies. For instance, tweaking the parameters of our hypothesis by our learning rule is equivalent to moving around in some optimization landscape. If we understand the geometry of this landscape better, we could pick an algorithm for optimization that could have an easier time reaching the lowest points of this landscape, where the statistical risk is minimum. Having an easier time means spending less time wandering around in the wrong places, such as saddle points or bad local minima, where the statistical risk could be much higher than what it could be. There are vastly many more issues to be studied regarding optimization too, such as sensitivity to the initial location in the landscape where we start our optimization at, properties of optimization trajectory such as mobility, the kinds of trajectories possible, and so on. These are all subjects of study in the optimization theory of learning.

Generalization

The problem of generalization is a direct manifestation of the problem with artificial rational agency, namely the artificial induction problem. Here we have a specific form of induction in the learning problem, in the sense that we're learning a function that generally knows more than just the specific examples we have already seen. Thus, this function approximation can be said to be arriving at a general rule. Yet, practically we only judge this generality by a test set. That is no proof for the "rule" to be true, and that is exactly why we study generalization, to ease that concern.

We are going to talk about such studies in the next section. Essentially the goal of a generalization theory is to predict the worst-case scenario for the error on a test set. That is to estimate the upper bound on the error regarding unseen examples. This upper bound is sometimes called the generalization error. Solving the generalization problem amounts to reducing generalization error, and almost anything in the learning process can affect generalization. That is why generalization sits at the core of the learning problem, gluing all different aspects of learning together.

Having said that, we typically divide up the error into two parts. One from our error in how the optimization problem is set up, the representation, and the optimization principle that is used (there is nothing that the optimization algorithm can do about this error). The other source of error is in the solution that the optimization algorithm finds to the problem we have set up, relative to the best it could have possibly found given the data we already have (for instance, if it performs the task correctly on all the examples it has already seen, this source of error would be zero). The former source of error is called excess risk and the latter is referred to as the empirical risk. Together they represent the limit on generalization risk/error.

Speaking of risk, we must note that the frequentist and Bayesian versions of learning have vastly different views on generalization. Simply put, frequentists work on one hypothesis at a time, searching after just one hypothesis that works the best, and that's how they formulate their statements (like we have in our discussion above). Bayesians, on the other hand, always work with a full set of hypotheses and weigh them by evidence, and prior beliefs. Thus, the optimization problems they set up are quite different than that of frequentists, and hence the generalization statements they make are also different.

Frequentists say let's take the limit of the number of examples to Infinity and work with the unconditional probability of data, instead of conditional probabilities like the probability of data given the current hypothesis or the probability of this current hypothesis being correct given the observed data. In the Bayesian view, one cannot afford to wait until everything is known, and must start using the observed data to review multiple hypotheses and weigh the risk of each, then calculate the full risk of all these hypotheses by what's known as the Bayesian decision theory.

Note that there is no contradiction between the two different views, they are just different approaches to estimate generalization. The difference is that the frequentist approach may be too optimistic (in assuming how much data we'll get to see), and too trusting (in the already seen data which may be corrupt or bad). These are philosophical issues between frequentists and Bayesians that are not going to get resolved anytime soon. Though, one can suspect that we may be able to make some progress on that debate if we take the task of defining intelligence more seriously!

Statistical Learning Theory

To judge a learning system's generalization we must choose a test set and use the performance on that set as sort of an ultimate benchmark of the effectiveness of learning. On its face, it sounds too empirical given the freedom in choosing the test set. How can this ever be a proper benchmark for learning? Proper means sufficiently low dependence on the choice of test-set, such that we could practically ignore the dependence. Statistical learning theory (SLT) attempts to provide some answers.

Before we get into it, let's note that SLT is still being developed and is not yet a complete theory. That means there are extremely relevant settings of the learning problem that the standard (textbook) SLT excludes. These new settings are currently under intense research, coming together in some extended theory endowed with a richer picture. Let's call this TBD theory, the extended-SLT, to distinguish it from SLT, the good old self-consistent theory. Let's cover SLT first, and that puts us in a position to discuss extended-SLT next.

Per our discussion of the formal learning problem, learning starts with choosing (no matter how trivial or complex) a representation and an optimization problem to minimize the statistical risk. However, we only ever have a finite set of samples, so direct optimization of the true risk is not an option. We need another optimization principle to substitute it with. This substitute is called the empirical risk minimization (ERM), which is the minimization of the average of loss function values over the finite set of samples we have seen (empirical risk of the hypothesis), whereas the true optimization would be the minimization of the average of loss values over an infinite set of samples (true risk of the hypothesis). The starting point of SLT is to show whether this could work at all. This substitution could work only when the sequence of ERM-based risks converges to the minimum true risk (as we see more data). This is known as the consistency condition and it is the bare minimum requirement of learning.

A key theorem of SLT proves that the necessary and sufficient conditions for consistency of learning empirically (ERM principle) are the same as the convergence of the empirical risk of the worst hypothesis in the space to the true risk (of that worst hypothesis). The worst hypothesis is the one whose empirical risk may deviate the most from its true risk, given our sequence of empirical examples. The key theorem suggests that we could just drop the dependence on hypothesis and try to have an analysis of learning and learnability that is uniform over different members of the hypothesis space.

By analysis of learning and learnability, we mean measuring the quality of learning. When can we say we are successfully learning? How close do we have to get to the true function we are approximating? Is there something we cannot learn? What would be non-learnable?

We could use the consistency condition as a notion of learnability, but that would be a very weak notion since it is only an asymptotic condition, it doesn't say anything about how fast we learn (how fast we converge to the true minimum risk). A consistent learning machine may be too slow to learn. In other words, it may require an impractically high sample complexity, defined as the number of examples needed to get usefully close to the minimum true risk.

Therefore, a good notion of learnability should consider sample complexity and require it to be sufficiently small, i.e. fast convergence/learning. But what should be considered fast exactly? To answer that, we should leverage intuitions from computational complexity theory in theoretical computer science — a field that analyses how fast different algorithms are as a function of their input size. There, polynomial dependencies are considered fast. Applying the same notion to learning, we can define learnability by demanding polynomial-speed convergence of ERM risks to minimum true risk. What does that mean? Say, we demand for the ERM risk to be away from the true risk by at most A amount with probability at least 1 minus B, then the amount samples we need in order to fulfil our demand can only grow as a polynomial function of A and B, not any faster. If this condition is satisfied, the learning problem is called PAC-learnable, where PAC stands for probably approximately correct. The variable A quantifies the approximation quality and since we are learning statistically, we need to measure the uncertainty in this approximation as well, which is captured by B.

Wait, how do you measure the uncertainty when we don't know the probability distribution of the examples? That's the whole point of learning, had we known the distribution, there wouldn't be any learning. So a very crucial assumption underlying PAC-learning is that there exists a fixed (that is, fixed over time, i.e. stationary) distribution that all examples come from and we will never get a sample from outside of that. That's the axiom here, and to question that, is currently only a work of philosophy, albeit a very important one which we have to come back to in the second volume of the book. Unfortunately, for now without this assumption we cannot make any statement about the generalization quality of learning in SLT.

Now the question is, when would this PAC-condition be satisfied? Enter the "fundamental" theorem of learning. It turns out that finiteness of a quantity known as the VC-dimension[89] which is only a property of the hypothesis space, not only makes ERM consistent independently of the probability distribution of examples, it also happens to be the necessary and sufficient condition for fast convergence (PAC-learnability) too.

Since the VC dimension is not a property of any given hypothesis in the space (class of functions), its finiteness means we can learn any function in that class with the same sample complexity in terms of the A, and B parameters of PAC-learning. This independence from individual hypotheses within the space, is called uniform learnability. That is the same as saying PAC-learning describes uniform learnability.

That still leaves us with the question of what happens if the VC dimension of some hypothesis space is not finite, can we still learn it? And how fast? Infinite VC dimension means it's not possible to learn uniformly. So what if we allowed the sample complexity to depend on the choice of the hypothesis within the space, to be a different polynomial function of A and B for each hypothesis? If this condition is satisfied we can still learn problems in spaces with infinite VC dimensions, except non-uniformly. Non-uniform learnability is therefore an extension of uniform-learnability (PAC).

You can think of non-uniform learning as some union of uniform learning problems. Though, now we can add in more of our prior knowledge or bias than just in the choice of the overall hypothesis space, which was our only freedom in a uniform-learning problem. We can now choose how we want to weigh each hypothesis subspace (the space of each uniformly-learnable problem) relative to other subspaces. Therefore, in non-uniform learning not only we want to minimize the empirical risk but also simultaneously optimize the choice of the subspace for learning[90] so as to achieve overall lower risk. This principle is known as the Structural Risk Minimization (SRM), an extension of ERM (for uniform learning) to non-uniform learning problems.

---

[89] Suppose any collection of D or less number of points in the input space with arbitrary labeling can be shattered (separated by a classifier) by at least one hypothesis from the space of hypotheses, then the VC-dimension of that space is the maximum value for D. Higher D requires more degrees of freedom from the hypothesis space to be able to accommodate shattering of arbitrary D points. So, VC-dimension is a measure of the complexity of the hypothesis space. While there are many other complexity measures, the fundamental theorem of learning is what makes the VC dimension special.

[90] This is also known as explicit regularization.

Every uniform-learnable problem is also a non-uniform learnable problem and every nonuniform-learnable problem is also a consistently-learnable problem. Therefore using this progression, you can see consistency, the weak notion we mentioned first, as a notion of learnability in which you allow the sample complexity to depend not only on A, B, and a particular hypothesis from the space but also a particular distribution of examples. That's why as shown by the key theorem for the consistency of learning, we only need to guarantee convergence of the worst hypothesis in the space.

To summarize the answer to what is not learnable, a problem could be not PAC-learnable, not non-uniform learnable, or not consistently learnable. Let's also note that we can always make learning consistent by simply memorizing every single example.

What about the opposite case, is there a learning algorithm that could learn everything well? There is a famous theorem in SLT called the no-free-lunch theorem which says for every learning algorithm if you fix the total number of examples it's allowed to see, there exists some distribution of examples and their labeling (corresponding to some precision task) that the algorithm would perform poorly on. That signifies the fact that we have to always supply some prior knowledge at least in the form of the choice for the hypothesis space, appropriate for the task of learning, i.e. the actual function we are aiming to approximate. If this choice is poor, even if we learn and get to the true risk, there is no guarantee that the true risk is small in this hypothesis space (compared to the lowest possible risk, called the Bayes optimal risk).

Prior to the no-free-lunch theorem, we only talked about notions in SLT that demand a certain number of examples to get to some desired quality of learning as measured by the true risk. What if we fixed the number of examples, say N, and just asked given N examples how well we are learning, that is how low is our true risk. In the previous section, we called this the generalization error and discussed that it is decomposable into two separate sources of error. One is the error in solving our empirical optimization problem i.e. empirical error (also known as estimation error) and the other is the left-over error from approximating the true problem with the empirical one, called the excess risk (also known as the approximation error).

Going forward, we'll refer to these two terms as empirical risk and excess risk which together amount to generalization error or true risk. Also, true risk refers to the frequentist risk of a specific hypothesis that our empirical learning picks.

Once we have a learning principle we can try to calculate an upper bound on the true risk (our measure of how well we are learning). This is mainly an exercise of bounding the excess risk using convergence and consistency requirements.[91] SLT is full of such theorems, each stating a generalization bound for some learning problem. However, all these bounds are qualitatively similar. By that I mean the excess risk can be shown to be smaller than a value, where this value depends only on some complexity measure of the hypothesis space (such as VC-dimension) and the number of examples seen, N.[92] This dependence is qualitatively simple: the higher the complexity of the model, the higher the excess risk bound, and the higher the number of examples, the lower the excess risk.

Obviously we want the lowest amount of excess risk, so that if we did a good job on empirical risk minimization, we would automatically be doing a good job on lowering the true risk too. If the ratio of complexity to the number of examples was large, low empirical risk would still result in high true risk. Based on the form of these generalization bounds, SLT universally recommends lowering the capacity (complexity) of the learning model (hypothesis space), as long as it doesn't hurt the empirical risk too much. The SRM principle tries to capture this intuition explicitly, setting up an explicit optimization to find a balance between underfitting (too little complexity) and overfitting (too much complexity). The balance point fixes an appropriate tradeoff between complexity and empirical risk.

Following this main intuition of SLT, we are advised to build smaller models (or smaller networks in the case of neural networks). Moreover, to control the capacity more rigorously, we should be inclined to prefer methods that allow for a straightforward control of capacity. This is strictly the case for the class of algorithms known as support vector machines (SVMs), where maximizing some margin, e.g. between the separating hypersurface and the closest example (support vectors) automatically gives us the smallest capacity (VC-dimension in this case) subset of the overall hypothesis space of SVMs.

SVMs became the poster child of ML in the 90s. They achieved remarkable practical success, especially via the use of the so-called kernel trick which allowed for injection of designable nonlinearity into a learning machine. It popularized kernelization in ML wherever possible. This was motivated not just by the flexible nonlinearity that kernels can bring in but also by other

---

[91] We can also derive generalization bounds using stability requirements (requiring changes in risk due to slight changes in the input to stay small, i.e. stable). Stability-driven results are qualitatively similar to consistency-driven bounds in terms of their convergence properties.

[92] The leading order typically goes as the square root of the complexity over N.

statistical learning theorems that make structural risk minimization a lot simpler in practice.[93] Most importantly, the SVMs and their success seemed to be totally explainable by SLT and this was a huge validation for SLT, which on its own, had given rise to very powerful theorems as we briefly mentioned above. What is the power of those theorems?

Do you recall the 3 theoretical aspects of learning, namely, representation, optimization and generalization? Well, SLT started (in the 60s and in the context of control theory) very ambitious and went straight after the core aspect, and the ultimate goal of learning, i.e. generalization, hoping we could find a rather universal theory that is independent of the details of the first two aspects of learning. And as we saw the results that SLT discovers are quite general, not specific to any particular representation or optimization algorithm. This is what makes it powerful. Now the question is whether it is also complete? Whether SLT is telling us the full story? Is it the best we can do if our goal is to achieve the lowest possible risk? Putting powerful theorems and validation by practical success stories together, you can see why people (until recently) used to take the silent assumptions behind SLT (and its recommendations) for granted and not ask such questions.

Extended Statistical Learning Theory

The answers to these questions are as follows. Maybe SLT could be considered complete but only within its assumption. However, we can do better if we step outside its working box. This is an empirical fact that deep learning has shown us.

Recall in our discussion of the upper limit on the excess risk. We mentioned that when complexity or capacity (C) of the model is larger than the number of examples (N) this upper limit on excess risk is too high and empirical risk minimization cannot guarantee any generalization. Therefore, SLT recommends lowering the capacity. However, an alternative and a more natural take-away would have been that this upper limit is too high to be of any use, and that instead we should look for a tight bound on generalization in cases where capacity is really large. SLT simply

---

[93] In particular, Representer theorems, which show we can represent the optimal solution of the empirical risk in these kernel machines as a function that is evaluated only at the set of input examples. If you're familiar with reproducing kernel Hilbert spaces, consider expansion of the optimal solution (minimizer of empirical risk) over the complete basis (reproducing kernel evaluated at every point in the space). Now, since the loss function only at each input example enters the empirical risk, only the subset of basis functions that are the evaluated kernel at input examples survives the expansion. That is essentially the statement that a representer theorem makes.

ignores high capacity and says look at the small capacity regime, in which case, it proves that we should refrain from getting to zero empirical risk without lowering capacity (what SRM does).

This was a silent shortcoming of SLT which is why it doesn't address the success of deep learning models, where the capacity is much larger than the number of examples (a condition known as overparameterization standing for having way more parameters than data points).

This is an important example of where SLT needs to be extended. There has been lots of recent attention paid to studying generalization when one moves from under-parametrized regime to overparameterized by incrementally increasing the capacity (C) of the model. SLT bounds tell us that increasing C beyond some locally optimal (e.g. SRM-chosen) C, would hurt generalization, until C is about N. However, it doesn't say anything as to what happens after that. A phenomena being tagged as "double descent", shows that after that initial increase of generalization error until the interpolation point (where we are fitting every seen examples, i.e. zero empirical risk, which is roughly at C=N depending on how we measure C), the error starts to decrease again with the increase of capacity, which totally breaks the wisdom of SLT, as well as just the basic intuition from statistics which says you need more data points than the number of free parameters you want to estimate using the data.

This phenomenon was first observed by a few physicists, studying statistical mechanics of generalization in the 90s.[94] It goes without saying that it was not taken seriously and hence not followed up on, until recent years where the unexpected (according to SLT) success of deep-learning has forced us to take it seriously. There are many things in deep-learning practice that we don't yet have a full theoretical understanding for. Many of those issues don't necessarily challenge SLT. However, there are some that do. Overparameterization without overfitting, does certainly and most explicitly challenge SLT.

Once you challenge SLT, you're challenging our understanding of any learning algorithm, not just deep neural networks. What I'm calling the extended-SLT here is just a placeholder for what is still being developed. Unlike SLT, we can no longer afford to ignore the details of the other two aspects of learning i.e. representation and optimization in studying generalization.

---

[94] "Statistical mechanics of learning: Generalization" Manfred Opper, 1995.
Opper M., Kinzel W. (1996) Statistical Mechanics of Generalization. In: Domany E., van Hemmen J.L., Schulten K. (eds) Models of Neural Networks III. Physics of Neural Networks. Springer, New York, NY.

Theory to Practice

The theories of learning we have been talking about only cover learning principles and some generalization guarantees to rely on. The last step of solving a learning problem is to construct a detailed algorithm guided by all that theory. This last part remains highly empirical. That involves constructing and testing an algorithm, iterating and testing again, on and on, with little to no theory to go off of. Even when it comes to generalization guarantees provided by the theory, the theoretical upper limits on the risk may be too high or far off the actual risk for the learning machine due to assumptions by the learning theory as to where the data is (and will be) coming from. Thus, those risk bounds in SLT are only proxies for actual performance.

At least for the time being, ML is a highly experimental science. Just like any other experimental science, the study is constrained by the feasibility and cost of the experiments. In the early days of ML's popularity, the 90s that is, compute resources were a lot lower than today's, and a lot more expensive, so were the ML experiments. When experimenting is expensive, one tends to follow whatever theory is available to lower the cost of blind trial and error. Going with more theoretically understood ideas can massively lower research risk. This is just a special case of the widely quoted principle that "There Is Nothing More Practical Than A Good Theory".[95] Vladimir Vapnik, the main figure behind the frequentist SLT (of last section) also starts his 1995 book on "the nature of statistical learning theory" with this quote.

On the contrary, if the experiments are feasible and their costs are less than that of building a good theory, people do almost all the experiments they can afford to do and only consult with theory when it starts to cost them unbearably too much resources. So to understand the theoretical progress in ML, we must pay attention to experimental feasibility and cost. In the past decade, experiments have become a lot cheaper. Regardless of what you consider a breakthrough in ML, almost all of them have been empirical and the theory is now playing a catch-up game to explain many successful experiments that aren't readily explained by existing theory. Our extended SLT of the previous section is only one avenue in theory work.

So far we have just been talking about the theoretical aspects of learning and the fact that we don't have enough theory for them especially for the first two aspects, representation, and optimization. But it's important to keep in mind that the practice of building learning machines and using them in the real world has many other aspects. Just like any other engineering task, it is

---

[95] The exact quote attributed to Kurt Lewin is "There is nothing as practical as a good theory"

full of design and implementation choices that current paradigms of learning theories don't even begin to talk about. Some example questions that one day could get inside the theoretical territory are: What you are to be learning exactly, towards what task specifically? What sets of tasks should you be jointly good at (what kinds of tasks could be bundled together)? How to recycle the learning machine for downstream tasks if any? How to maintain the machine? For now, we can only try and follow guesswork.

Remember in the formalization of the learning problem we unified all different learning types. We could do so because there were sufficient commonalities for a type-independent study of generalization guarantees. In practice, however, we need to bring in the differences between learning types into focus and try to balance the relevant trade-offs, of which there are many. In the interest of our conceptual discussion, let's consider a broad example.

Suppose we want to use a learning system towards not just the primary task it's learning about, but other yet-unknown downstream tasks. This choice involves balancing a tradeoff between "learning and inference". Wait, isn't learning a form of inference? Yes, any learning is generally a form of inference, however, as a technical term, inference is typically referred to a process that is post learning and independent of it (with learning being just a preparation for inference). In ML literature, sometimes they are distinguished with learning referring to training for a primary task at hand and inference referring to learning about some latent variables that explain the data and some structure behind it. Correct inference of latent variables could be broadly useful for other tasks (to be transferred for other learning purposes i.e. "transfer learning"). Often there exists a tradeoff between being super good at learning for a current task (learning) and being prepared well for other tasks down the line that rely on the same domain of data (inference).[96] How one balances off this tradeoff depends largely on the design goals that are set for the learning machine. Are we after a one-time quick and narrow problem solving, or are we after a more adaptive slightly more open-ended learning machine? This question and similarly many others (including how to best balance their corresponding trade-offs) are not yet the subject of any rigorous theory but only of the art of design and engineering dictated by silent philosophies.

---

[96] This tradeoff between learning and inference is of a statistical nature. Not to be confused with the more computational tradeoffs toward a single task. For an explicit example see "Trading-off Learning and Inference in Deep Latent Variable Models" Lévy and Ermon (2018).

Different Camps

In the absence of comprehensive theories, it shouldn't be a surprise that different camps exist within ML practice. None is exempt from the commonalities of the learning problem, covered so far, but they do differ on their ideologies and approach towards building ML systems, ranging from what ML systems to build, what aspects and components of the learning machine to put more emphasis on, to what learning principle should sit at the core of an ML system. Should we welcome the existence of multiple different approaches to learning? Should we just figure out which one is the best and stick to it religiously?

First off, let's note that the no-free-lunch theorem (see the statistical learning theory section) has nothing to do with the existence of different camps. That's because a camp doesn't represent a specific algorithm, it's an only ideology and approach to building one. So that theorem alone doesn't prevent one camp from being better than all other camps at almost everything.

In his book, "Master Algorithm", Pedro Domingos, a distinguished ML researcher, divides up the ML landscape into 5 camps:

1. The symbolic camp. Any approach that uses logical inference on symbols. e.g. Inductive logic programming.

2. The evolutionary camp. Any approach that mimics the process of evolution, that is inter-life learning, e.g. genetic programming.

3. The neural network camp. Flagship example being end-to-end[97] training of a neural network with stochastic gradient descent.

4. The Bayesian learning camp. Aiming to learn the posterior probability above anything else and reasoning from there.[98]

---

[97] End-to-end means traditional pre-processing and post-processing stages of traditional ML systems are included in the much bigger model which maps raw inputs to the ultimate target outputs. For an extreme example, think of going from the raw images of the surroundings of a self-driving vehicle to the degree of turning for the wheel, or push on acceleration or brakes.

[98] Though, considering Bayesian learning a separate camp on its own is a bit strange. The Baysian approach is an approach to probability not learning. Any learning method can have a Bayesian version that is more robust at least in principle, compared to its frequentist version.

5. Analogizers camp. Approaches that make predictions exclusively based on similar examples. Such as support vector machines (SVM) and nearest neighbor algorithms.[99]

He essentially asks, given that different methods have different strengths in different areas, wouldn't it be great if we instead had just one algorithm endowed with the strengths of all different approaches i.e. a master algorithm? As we said, this isn't in violation of the no-free-lunch theorem. It's about combining the strength of these different methods into one which would still be subject to the no-free-lunch theorem although in an irrelevant way because such a master algorithm is, by definition, about doing the best we can possibly do. However, the author suggests that the master algorithm is some transmutation of these different camps as opposed to a fundamentally different approach to learning. Therefore it is important for us to cover the topic of combining approaches. Before discussing the combination of different approaches to learning, we must understand the combination of learners within any given approach to learning, which is a huge part of the practice of ML anyway.

Combining Similar Learners

Let's start with Occam's razor, a label for the bias towards preferring simpler solutions to the learning problem. We will have to revisit Occam's razor in much greater detail when we face uprooting simplicity in the second volume of the book. For now, let's regard it as a bias in inductive reasoning to choose, say, among the hypotheses that explain the data[100], the simplest one! The one with the least complexity. You can also interpret this as increasing smoothness of the learned function or minimizing any variation or singularity that is not strictly necessary for fitting the examples (training data).

To increase such smoothness, we saw that SRM would put an explicit penalty on the complexity of the hypothesis. We call that an explicit regularization. There are also implicit ways to

---

[99] Putting SVMs and nearest neighbour algorithms in the same camp may also be a stretch. In practice, they are very different, in almost every aspect, except for the explicit usage of similarity. SVMs only need to memorize the examples near the decision boundary (the support vectors) whereas nearest-neighbour needs to memorize all examples and in return, given enough data, they are at worst only twice as error-prone as the best imaginable (Bayes-optimal) classier (as proved by Tom Cover and Peter Hart, 1967).

[100] In the modern world of ML, we can think of explaining the data as the act of interpolating the data points.

increase smoothness which are all empirical methods as they all depend on the (training) data (and the performance on it). In these methods, there's no actual weight or penalty assigned to different hypotheses independent of the training data.

There are exactly two implicit ways to increase smoothness. One is through the optimization procedure. By not getting too greedy on the optimization work and not getting too carried away by the ERM problem we set up, we can increase smoothness. Using stochasticity is an example of non-greedy optimization. And stopping the optimization process (early) before it results in too much variation in the learned function, is an example of not getting too carried away by just empirical risk minimization. Stochastic gradient descent with early-stopping can do marvels as they do in modern neural networks.

The other way to increase smoothness is by averaging multiple learners. Any averaging can increase the smoothness as long as the learned functions are diverse enough. Combining multiple learners to generate a collective hypothesis is known as ensemble learning.[101] As you can imagine, based on our discussion on non-uniform learning, there are many ways to go about such collective learning. The simplest way would be to make predictions using majority voting among learners. However, voting or a linear combination (weighted voting) of learners may not necessarily yield any improvement.

To guarantee improvement we must make sure different learners in the collection can correct each other's mistakes, rather than all being prone to the same mistakes. In other words, they need to be diverse experts i.e. each be good at something and collectively good at different things. Even then, we may want to have a separate learner that learns how to best combine them — this approach to combining learners is known as *stacking*. Further, we can make the combiner unit depend on the input that each learner receives so that it can route the prediction task to the more suitable learner (or learners) for that input — this is known as *gating*.

The question is how to generate a collection of experts that are diverse? Suppose we have picked our approach and the representation form of learning for all learners. That is, let's ignore algorithmic diversity and focus on having the learners learn different things so that we have complementary learners and can gain something from combining them. For most of the history of ML, the way to do that has been via implementing an explicit way to ensure diversity. Here are the four main methods to do that:

---

[101] Ensemble learning usually only refers to voting-based algorithms, whereas what we are going to discuss here is quite broader.

1) *Bagging*, where we repeatedly sample a new training set from the set of training examples to train the learner. All training sets have the same size but the redundancies within them are varied — any given example may show up many times in a training set or not even once. Each time you get a new learner by training the same learner on the new set. The collective learner is a simple average of all learners. "Random forests" use bagging and are among the most popular ML algorithms of all time ever since they were introduced about 20 years ago (just a few years after the bagging technique was invented).[102]

2) *Boosting*. We can modify the training set but in a more controlled way than random sampling (as in bagging). In boosting, we generate learners in a sequence, each training set depends on the performance of the previous learner. Examples that are weakly-learned, by the current learner, will get a higher weight in the training of the next learner, and so on. The idea behind boosting goes even deeper than that: it's about asking whether or not it is possible to combine many weak learners to obtain a strong learner. In the early 90s, it was proven that in binary classification as long as weak learners consistently do slightly better than random guessing, boosting can increase the performance arbitrarily high, i.e. create a strong learner. What's surprising (based on SLT) is that boosting can continue to increase generalization even after training error is zero (zero empirical risk). That means that boosting is resistant to overfitting.

3) *Collection of local experts*. The idea is to use adaptive basis functions (learned features) that tend to become experts at different parts (different locations) of the input space. We can force them to be so by encouraging them to minimize their overlap in what's known as competitive (or winner-take-all) learning. "Mixture of experts using radial basis functions" is the flagship algorithm here — it is an example of gating that we mentioned above.

The problem with local basis functions is that we may need too many of them, especially in high dimensions. What if, instead of local representations, where each basis function is learning only one distinct location in the input space (as if we are paving the input space with these local basis functions), we used "distributed representations", where each basis function, or learned feature, gets activated by a whole region (consisting of many locations) in the input space? We'll discuss distributed representations in more depth in the next chapter as it is a foundational concept in ML. For our discussion here, we can look at it as

---

[102] Random forests also use another technique known as "feature bagging" to diversify learning about input features and that too is critical to their success.

what allows every input example to activate many different basis functions simultaneously, and in that sense, its representation is distributed! However, these regions (being large) will be overlapping, so we'll still have work to do to ensure that the experts are diverse enough. One (laborious) way would be to manually design them to be so, e.g. design explicit binary codes for each basis function such that the overlap of the codes (their *"Hamming" distance*) is minimized.

4) *"Error-correcting output codes"* do just that, in the context of multi-class classification. For each class, we can create a binary code of length K with K learners. Each learner focuses on a binary classification. We can ensure diversification among the learners by minimizing the hamming distance between the codes assigned to each class. Although it's been shown that even random non-optimal codes can do just as good, the problem with this method is that every basis function (in this case, every binary classification) can be easily learned since we have defined its output code a priori without any learning.

All these different averaging techniques, mentioned above, do improve smoothness and generalization when they are used in suitable settings. Although, they often come at the cost of losing any natural interpretability that base learners may possess. Let's discuss how they became popular.

Remember that ML used to exist even in the 60s, though the computing requirements were too high for it to take off. Also "AI", which was already a popular pursuit, went through its first "winter" in the 70s, in a sense, also because of low computing power at the time. Available computing power grew significantly for the 90s machines, but it was still very low compared to today's standards, so these averaging techniques and ensemble methods (right after SVMs and kernel machines) were among the biggest contributors to extend the success of (frequentist) ML in practice, for years to come (in the late 90s and 2000s).[103]

A lot of these averaging techniques may seem like reinventions of Bayesian learning ideas from statistics, except that they remain frequentist and are not equivalent to Bayesian model averaging.

---

[103] As we mentioned in our discussion of pre-modern ML and the fall experts systems, probabilistic models became a center of intense research in the 1990s bridging statistics and computer science. Bayesian methods took over and replaced expert systems. However, internet companies in the early 2000s used much more frequentist ML than Bayesian algorithms (which are more sophisticated and typically require more computation in return for more robust decision making). Bayesian learning continued to be the hottest edge of ML in the 2000s (especially in unsupervised learning with algorithms such as Latent Dirichlet Allocation).

In fact, one can construct the Bayesian versions of each of these frequentist techniques we've mentioned here.

Combining Approaches

So far we have only talked about averaging similar base-learners. What if we want to diversify algorithmically, and combine approaches in different ML camps to get better than any individual camp (and maybe even find the master algorithm)? We should find a way to hybridize them.

Suppose we break the learning problem into multiple stages, where each stage can be considered to be fulfilling a separate task although all stages are implicitly working towards a common end goal. For instance, pre-processing the data with some objective, finding a good representation with another objective, and using the representation towards some ultimate objective, would constitute a 3-stage pipeline that can be broken down even further. What happens if we turn any given stage into an ensemble learning where different learners from different approaches are just treated as different members of the ensemble?

Ignoring basic challenges around mismatches in representation, data structures, and "calibration" differences, we still cannot quite merge different approaches towards any given task and expect any improvement. Of course, we can just have them vote and weigh their votes based on their performance on some "cross-validation set" (a test set that you're allowed to see while learning), but since they are completely different from each other, we would have to consider a non-uniform prior belief over different learners. The prior belief serves to supplement pieces of knowledge that are not captured by a simple performance score on a cross-validation set, such as our own prior knowledge about which learners are supposed to work better or are more suited for the situation.

Specifying such a prior is a tough task. If we have such a non-uniform prior, it can already tell us which algorithm is supposed to work best for the task at hand, and then we are better off just using that one, instead of trying hard to combine approaches. On the other hand, if we, for a second, ignore the differences between the suitability of different approaches (that is, to put a uniform prior over them), we know that in reality one of them will turn out to be outperforming all others. Given that none of them is a weak learner[104], combining them is only worsening the strongest of them.

---

[104] Candidates from every camp are already strong learners. If not, then we can always have them form their own ensemble of similar weak learners using the techniques of the previous section, before hybridizing with other approaches

Therefore, we are better off without any hybridization of approaches on the same stage (subtask) of a learning problem.

In a more general setting where the task of a given stage is separable (decomposable) to several tasks (say based on the region of input space, similar to the case for "mixture of experts" we mentioned earlier), we may benefit from a parallel architecture, where different subtasks (input types of regions) are handled by different approaches. Yet, for that to work well two conditions must be satisfied. One, every approach must be exclusively better than all others in its own subtask. Second, we have to be able to properly learn some "gating box/function" that is supposed to take care of this distribution among learners from different camps/approaches.

These special conditions can very often be satisfied, but it may still be quite a laborious design and engineering task to build such a hybrid system towards only one end-goal. In other words, the lack of generality of architectures for combining approaches is what diminishes their appeal. That's why general designs are pursued in what's known as "cognitive architectures", where the hope is to design and build platforms that could support "general intelligence", and maybe host a "master" algorithm.

Having said that, we can still gain rather easily from combining approaches in a *serial* manner, where different approaches are used at different stages of the pipeline. A successful example of hybridization can be in bringing the representation from a deep neural network and feeding it to an SVM or a nearest neighbor algorithm for some classification or recommendation task down the line.

Chapter 6

AI as Deep Learning

Based on our discussion of combining Learners and/or approaches, it's clear that if we knew how to best break the learning problem into smaller subproblems where each subproblem would be much simpler than the original problem, we could then successfully apply our combining techniques and keep improving performance. However, this is a big if and indeed the real challenge. Consider error-correcting-output-codes (ECOC), discussed in chapter 5. As we mentioned, the main challenge there is with how to ensure that the individual learning tasks would be actually easy, given that we have to design the output codes manually and prior to any training. What if to overcome this challenge, we could use the data and the learning machine itself, to figure out how to break the problem into pieces optimally? Well, that's exactly what deep learning tries to do. Not only can we automate ECOCs, we can also figure out the optimal (intermediate and output) codes implicitly.

Deep Learning as Decomposition Learning

Before getting into more details about the how. Let's adopt a more fundamental perspective. And that's the perspective from the problem of *decision-making*.

A Generalized View of Decision Making

In a broad sense, decision-making is perhaps all that we do, or any AI agent does. Any learning task can be thought of as a special kind of decision-making. Setting a variable is a decision to pick a value among several possibilities. So learning the function f from data, instead of some other function g can be viewed as a decision, and thereby all types and settings of learning problems can also be viewed as decision-making problems. Be it, say, setting all control variables in a manufacturing plant, or classification of a data point. Both can be regarded as taking actions toward some utility maximization, and both admit analyses by (at least a Bayesian) decision theory.

Now suppose you want to break any of these decision-making tasks down to smaller subtasks just like you would do in an organization (in our human world). In particular, consider a hierarchical organization whose overall performance is being represented by a chief executive. Each direct report to the chief executive, after some initial scrambling phase among the peers, picks up some unique role. Each roll then comes with some implicit utility to the maximized corresponding to its implicit objective and therefore is confronted with its own set of decisions to make. While all this is ultimately in service of the overall objective of the organization, as we go down the hierarchy, the problem keeps getting broken down to lower level subproblems, lower-level objectives, and similarly, the decision-making subproblems shrink in scope.

Now suppose the top-level decision making is just a go or no-go decision by the executive, i.e. a single binary choice. Further, suppose all the sub-decisions are also of the same kind, that is, some binary choice, all the way to the lowest level. Outputs of each node in the organizational network flow up to be integrated in a higher-level decision making — this is the bottom-up flow. There is also the top-down flow, where the overall output of the organization needs to be corrected (say, based on some environmental feedback, e.g. market, investors, etc.). In the top-down act of adjustment or correction, not everything needs to be modified, only some subdivisions and sub-decisions need to be modified. To figure exactly which ones need to be modified, given the higher-level feedback, is the same problem of "credit assignment" which we talked about in discussing the optimization aspect of learning.

This breaking down of decision-making problems in a network is an immensely general perspective. It allows us to understand deep learning architectures as special cases of something more general! In the example above, we considered the simplest possible type of decisions, binary decisions (or outputting a number between 0 and 1, or -1 to 1, etc.), and essentially arrived at what's known as *feedforward* neural network architectures.

If you consider any connection between peers, sending your output/report to a member in the same level of the hierarchy, you'll get what's known as a *recurrent* network. Any intra-layer connection results in a recurrent network. Your decisions tomorrow would use the output of your today's decisions. You can consider this a connection between you and a copy of you in the same layer of the hierarchy, as an intra-layer connection in a recurrent network of your states of being.[105]

Decomposition Learning

So if you ask what's really new or special in deep learning, that sets it apart from other ML algorithms, the answer is that you are *learning* the proper decomposition of tasks, and how to combine multiple learners. Thus, a better name for deep learning, as suggested by our discussion above, would perhaps be "decomposition learning", a problem of fundamental and universal importance, which is not fully solved yet!

We define decomposition learning as learning to recursively break a task down into smaller, as well as easier, subtasks (again, we can interpret all tasks and subtasks as decision-making tasks) until we reach either trivial subtasks or ones we have already solved/know how to perform. To see the fundamental importance of solving decomposition learning, it suffices to say that solutions to it would enable automation of most enterprises especially on the management and executive side! Deep learning is a crucial part and a special case of decomposition learning and will always be. Why?

Again, decomposition learning is about learning to break a task into smaller and easier tasks. To explicitly see how that is a generalization of deep learning, let's interpret "smaller" as involving fewer computational units, and interpret "easier" as requiring fewer examples (for learning to generalize practically well). Finally, as we already mentioned, we should interpret "learning task breakdown" as automating the task of breaking down, rather than doing it manually the way ECOCs, or other learning-combo techniques, do. The ONLY non-superficial thing that sets deep learning apart from other ML algorithms is that it's the only approach that does decomposition learning as defined above. How?

Deep learning's approach to decomposition learning involves two main aspects:

---

[105] In this language, one can see that any decision that has to be made by the same decision maker more than once, causes the whole network to be recurrent.

1. Learning subtasks using *width* i.e. a layer made of more than one artificial neuron. (We'll discuss what it means below, but essentially, width allows for non-local learning — in ML, non-local simply means, not learning about one location (in input space) at a time)

2. Keep breaking subtasks into further subtasks using *depth* i.e. having more than three of these layers. (As we'll see, depth allows for learning about non-smoothness[106] without introducing a tradeoff with non-local learning)

Decomposition learning by definition involves task breakdown, so any approach to it will involve some network with multiple units integrated in some architecture. Width and depth are the most trivial features of any architecture with a flow from some input set to some output set. Therefore by themselves, depth and width don't say much about whether decomposition learning is happening or not. So when are they exactly in service of the compositional learning that deep learning aims to perform?

Depth

Depth is only relevant to decomposition learning when we cannot achieve the same function-approximation using a less-deep (less number of layers) representation, without blowing up the width of any layer exponentially. When learning is actually leveraging depth (i.e. truly deep-learning), the function that is being approximated admits compositional learning. Going reverse, one way to make such a function is through repeated compositions. No matter how abstract the domain, or how complex the patterns created by a function that you wish to approximate using examples, if you can imagine any form of building it by repeated compositions, that function can in principle be represented by a deep network.

Nature is full of such functions. At least all concepts familiar to us humans are a subset of such functions. To understand how ubiquitous these functions are, and how significant is the ability to learn them well, think of every scenario where one can ask the question, "what is this or that made of?"; any time we believe in the concept of *ingredient*; any time we believe we can *compose* something; any time we build something by putting (usually smaller) things together whether in the physical or conceptual world; anywhere we believe in a *reductionist* explanation, which is almost everywhere in our world because that's how we think, and how we are taught! Note that this extends

---

[106] In ML, non-smooth often just means a widely-varying function as in with a large Hessian component. But to be more precise for mathematicians, we are just talking about Lipschitz-smoothness.

to the temporal domain as well, where Markov chains (dynamics behind almost all everyday-life phenomena) made of simple steps can bring about complexities that are too hard to unravel without decomposition. To efficiently learn functions arising from any of these cases, we need a representation with a depth necessary to undo the composition, that is, to do decomposition learning.

Most complexities that we humans care about, and certainly the ones we create, are the products of repeated compositions. By this process, we create interesting complexities, we create exponentially (exponential in the number of ingredients that are put together in the generation process) many different patterns, situations, or instances.[107] If we ignore such compositional processes behind the apparent complexity, understanding the complex pattern, and telling apart different instances of it, would require an effort proportional to the number of ways it could vary (again, exponential in the number of generating ingredients). That is the origin of the exponential width-blow-up if we insist on getting the same performance with a shallow architecture as we would with a deep one. Thus, having depth means leveraging the compositional structure present in the pattern we are trying to learn and predict.

Theorists know that whenever there is an exponential somewhere, there also exists a *logarithm* function that tames that exponential. One just has to look for a process that takes the logarithm of the exponential (the combinatorial explosion). If these exponentials are nested, which are in the case of repeated compositions, the logarithm-taking process has to also be applied repeatedly to unwind the complex pattern into something simple. The "depth" in deep learning algorithms is effectively that repeated logarithm-taking process.

Non-Locality

Let's get a bit more concrete on how logarithm-taking takes place. But before discussing the how, let's first see why we need to have some effective logarithm-taking even when there are NO repeated compositions. The reason for that is known as the "curse of dimensionality". Just like overfitting, the curse of dimensionality is also a general challenge in any learning problem, except it is only relevant for a high number of dimensions (of the input space). High means more than a few, say

---

[107] Note that there is no explicit exponential function here. We are just referring to "combinatorial explosions" for the number of possibilities.

more than 5 to 10 dimensions. That's almost always the case, and therefore all non-trivial problems of interest in ML suffer from the curse of dimensionality. What is it?

Remember learning is about approximating some function. Suppose you want to approximate the value of a function at some point x (in the input space that we are assuming to be high dimensional), using some examples "near" it. The problem is with the meaning of "near" in high dimensions. Any region you imagine, centered around point x, would have almost all its volume composed of the points near the edge of the region. We don't have a direct sense of this because we live in a low-dimensional space, namely our 3D space.

It's easy to recognize that, mathematically, volume scales with radius to the power of the number of dimensions. So, consider a sphere of radius R in, say, D dimensions, and another sphere inside it of a smaller radius, r. The small sphere occupies only r divided by R (a number smaller than 1) to the power D. If r is half of R and we are in 20 dimensions the volume of the smaller sphere is a MILLION times smaller than the sphere with just twice the radius. So that means the number of ways to be close to point x is way too many, and we need exponentially more examples to reliably estimate the value at x. This is the exponential that we need to take the logarithm of. Since as you just saw, even without repeated compositions, just with 20 different dimensions where along each dimension we have a binary variable (representing whether, along that dimension, the example is inside a smaller sphere, or outside of it in the bigger sphere), you get a million different possibilities.

Now, how do we take the logarithm? The problem arises from high dimensionality, the fact that we are looking at the whole input space where all the features (dimensions) are present, and they can combine in exponentially many different ways to confuse us (we being the learning algorithm). Then to figure out the value at x, we have no choice but to get exponentially close to the point (with exponentially more examples) and this is a *local* approach. To get away with fewer examples we need a non-local approach as you'll see. We need to work in subspaces of the input space (not the full input space at once), and with one feature at a time (not all features at once). That is effectively breaking the learning task into multiple subtasks, and in this case, it is known as a distributed representation.

Let's unpack that using our example above where D=20 dimensions and each feature is a binary variable. That gives 2 to the power D possibilities, which would be a million, different cases to tell apart, and we'd need a similarly large number of examples to do that well. But if we break the problem into 20 different subproblems wherein each subproblem we only ask whether two cases are the same or different according to one binary feature (for instance as represented by one artificial neuron), we would need about 2 times 20 examples to be able to learn about the million cases. That is the logarithm-taking in action. When we learn about binary features independently they can each slice the whole input space into 2 pieces for me, and each one does it in a different way. Now collectively they are representing 2 to the power 20, that is, a million different regions.

This is called a distributed representation and is nothing new. Humans do it explicitly all the time. When we learn to classify regions by north and south, and separately by west and east. This is breaking down the problem of classifying four different directions on a 2-dimensional(2D) plane into 2 subproblems. When we say north-east or south-west, we are using a distributed representation. A non-distributed representation would be to label the 4 possibilities by 1, 2, 3, and 4 instead (for northwest, northeast, southwest, and southeast). The other example is the ECOCs, as we mentioned earlier.

In these examples, we are designing a distributed representation by hand. The downside is that a subproblem in which the dimensions may be lower than the original problem, may still not be easy. So just using each dimension of the full input space (e.g. original features that come with the data) to define a subproblem may not work. However, if we allow for each dimension to rotate freely and then allow for some non-linear transformations of them, we can find subproblems that are easier to solve. And we don't want to do this manually. Deep learning's secret is to combine the power of distributed representations (width) with depth as we discussed above to automate this process and figure out which way the problem breaks down best. That is, it can automate finding easier subproblems.

Finally, in what sense is this a non-local approach? It's true that fundamentally solving the learning problem using examples must use at least some notion of similarity, closeness, or analogy, between what we want to learn about and the examples to learn from, otherwise no learning is possible. But the use of similarity or analogy in the "analogizer camp" (such as the nearest-neighbor approach or kernel machines), mentioned in chapter 5, is a much stronger one. Strong in the sense that, there, we use only one notion of similarity in the full input space where all the examples live. We can call that a global-gauge-of-similarity, since by using such notions of similarity, we can judge how close or far two points or two examples are from each other, regardless of where they happen to be in the input space.

Contrast that with, say, a multi-gauge-similarity where you break the original learning problem down to several subproblems, and each one effectively introduces its own measure of similarity. In this case, because we are focusing on one subproblem or feature at a time, there is no guarantee that the examples we are using to learn about point x in a subproblem, are actually close to point x in the full input space (as measured say by some corresponding global-gauge-of-similarity). As you can see in the figure below, even though the example is far from point x in the full input space (the 2D plane in this case) we can still use it to learn about x in a subproblem which is defined by the projection of the space along the horizontal axis. This is why this approach can be called a "non-local" approach. In contrast, all out-of-the-box approaches in the "analogizer camp" are based on some global similarity measure, and therefore can only learn locally, and doing so, as we saw earlier, is prone to the curse of dimensionality.

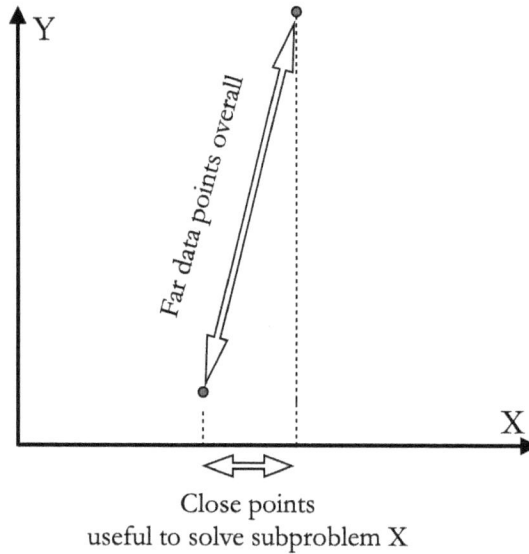

Close points
useful to solve subproblem X

Can we fix up the local approaches by using several similarity measures, say, using multiple kernels based on which region we are at in the input space (decided by a decision tree for instance)? Again, no, we would still need to break the space into exponentially many different regions, and even then it's not clear what similarity measure for that location would compensate for the ineffectiveness of the close-by examples we are using to learn in that location. The choice of measure is not the fundamental problem here. With any given measure to be used globally, at each location, we have different examples that act as different templates to measure the similarity against. For that reason, regardless of the similarity measure used, similarity-based local techniques are often referred to as template-matching techniques.

What that tells us is that dividing up the problem is not equivalent to decomposing it. In the division of the problem into multiple pieces (regions of the input space), we are still solving the original problem, albeit smaller versions of it. That shouldn't count as breaking the problem down into subproblems. Alternatively, we could say, breaking down the problem into subproblems that have nothing to do with each other is not (the way to go about) decomposition learning.

We have been saying that decomposition learning is about breaking the problem into subproblems that are easy to solve. A critical part that is enabled by non-local learning (distributed representations) is that subproblems are not isolated from one another, what we do to solve one of them gets reused in the others, as needed. This sharing of knowledge (implemented by parameter sharing in deep learning) is what makes subproblems easier. That is, it allows us to learn with fewer examples.

Human Expert Labor

Having said that, one could potentially overcome the shortcomings of naive global-similarity (or local) approaches by working really hard and leveraging expert knowledge in the domain of the examples, which would involve designing many features and many similarity kernels. But even then, it would work out for just that specific problem. A different problem would require similar labor and a start from scratch.

Therefore, as we said, what's remarkable about deep learning is the automation of this process, which is an essential element of "decomposition learning", learning a breakdown of the full problem. And that gives us another perspective, which is that deep learning effectively learns similarity kernels without expert knowledge. This quality has enabled deep learning algorithms to take decades of work in designing kernels and engineering features for images and the whole field of computer vision by a tsunami. The same thing has happened to the fields of speech recognition, and natural language processing.

Even though the technique (deep learning) we just described is not specific to any particular domain, it doesn't mean that we can use it readily to do better in every problem and every domain. There are problems for which we have very little data but a lot of prior human or expert knowledge to leverage. In these cases, a deep network may perform very poorly compared to a "shallow" or "local" technique. Compared to, say, a set of nearest-neighbor learners with good similarity measures carefully tailored to the target context using decision trees, or a Bayesian network that can absorb a lot of our prior knowledge in the structure of the problem (including knowledge of all semantics and causal relationships in the problem). Both are examples that can vastly outperform a deep neural network in a very specific task where little data but lots of expert knowledge is available.

Not only the performance can be better but we could get other benefits too. One is interpretability. Remember just like combining learners loses interpretability, deep networks lack a natural interpretability. Second, in the Bayesian approach, you can correctly quantify your confidence in the results too, which you cannot in deep learning as it is practiced. Third, with these "non-deep" methods, we can also make a lot of progress relatively quickly when we just care about our narrow and isolated problem (as long as we can work hard to bring in human ingenuity and expert knowledge together). For these reasons, kernel methods and Bayesian methods were very popular in the 90s and 2000s. And they are still very appealing in the cases we just mentioned. The enterprise world is full of such problems, low with data, with a lot of domain-specific expert (and institutional) knowledge, and business problems that may be transient, for instance in cases where one pivots a whole business line.

So there is a tradeoff between how isolated and quickly you want to learn using expert knowledge, and how high a performance you'd like to achieve without expert knowledge. In the former, you pay with human expertise to handle a specific target application, which doesn't scale to all the applications we'd like to build. In the latter, you need a lot of data and sometimes innovations to collect, and label all that data, which may not scale to all the problems where collecting the data is simply too expensive or too challenging. Deep learning is after very high performance in "hard" learning tasks where codifying human knowledge is almost hopeless. In return, these deep learning methods are very data-hungry.[108] There is a saying in deep learning that "life begins at a billion examples"!

Wait a second, didn't we say deep methods need less data compared to local methods because they overcome the curse of dimensionality? Yes, but that's the thing, deep methods shine when we are after very high performance without any expert knowledge. If we don't use prior expert knowledge and demand the same high performance, local methods would need exponentially more examples, in other words, they won't work. But as we increase the amount of data available beyond a certain threshold, suddenly you have enough data to properly break the learning problem into subproblems and still learn each one well. It is in this regime that, when we demand a very high performance (in particular, generalizability in the task of prediction), a deep learning approach beats

---

[108] The upside is that there may be tasks that share a lot of commonalities, where we can reuse the solution to one problem to solve the other tasks which come with a very modest number of labeled examples. This is the subject of multi-task learning.

other approaches. The reason for that high performance is simply and solely the decomposition learning that is taking place in deep learning.

To conclude concisely, here's what we have said for far: non-local learning is more responsible for lower sample-complexity (needing less data), making the subproblems "easier", while depth is more responsible for more efficient representation (less compute units/neurons), making the subproblems "smaller", per our interpretation of deep learning as decomposition learning. It's therefore important to note the only thing that sets deep learning apart from other methods is that it is the only (special) case of decomposition learning we know of so far. Yet, there are many things it doesn't say about how to decompose, when you are tasked with many things and when everything is related to everything else, such as is in real life. Such a perspective seems useful in order to improve or step beyond deep learning.

Getting a Feel for What Deep Learning Actually Does

We just said, deep learning allows us to break the task into subproblems, which are different, in nature, from the full target task/problem, and allows them to share knowledge and recycle learning amongst each other. None of the shallow learning ones learn anything other than the target task and they want to go at it in one shot, though, could be in pieces like in local template-matching cases. That may all seem a bit too abstract, let's make it a bit more intuitive.

Consider a college student. Normally, s/he would be taking multiple courses, study them separately, and then take exams on them separately. That is because we have already broken down our knowledge into different subjects and topics with appropriate hierarchies, and the college has designed the curriculums for the students to follow. Now suppose there were no separate subjects or categories whatsoever, the lessons were coming at the student at random, the student would have to effectively switch between different topics in each lesson, there were no intermediate quizzes/exams/projects to build up the knowledge, and there were no exams but just one that is a mix of all exams held in just one session, in one day! At the very least, you'd expect the student to be confused, right? How much practice would one need to pass such an exam? "exponentially" more than in the normal case we are familiar with, right?

Well, that's exactly what happens in shallow/non-deep networks or local methods. The models are confused and for them to perform well, they need a lot more practice to perform well. That's why if we insist on their shallowness, their width has to explode, i.e. they need to become too wide, per our earlier discussion on depth.

In this hypothetical example, it's clear what the best strategy for the student is. Given that all the lessons are given interchangeably and the exam overlaps all the concepts, the student has to break the lessons into multiple subjects, learn the concepts separately, and then combine them towards answering the questions in the mixed-up final exam. That's what deep learning does, reversing that mix-up.

How? Consider the space of those lessons/examples given in the mixed-up class. What deep learning allows us to do is to *retile* that space in many different ways. Say your space is initially tiled with rectangular tiles, we can first retile with other rectangular tiles that are of different size and orientation, and then we can curve them up and change the shape of the tiles using some non-linear transformation. And we can keep doing that. What this repeated retiling does is that in each one some set of examples/lessons that were far separated in the original tiling, now either are within the same tile or have fewer tiles in between. So each tiling could be useful to learn a different concept, the concept that is shared in common between those lessons that are brought closer together in a given tiling.

That's what deep learning tries to do: keep retiling the space until points/examples of different classes are "linearly separable". Shallow methods, on the other hand, try to do this in one step. A shallow method attempts to take the final mixed-up exam in one go, and there aren't enough practice materials (number of diverse examples) or resources (compute units) for a shallow method to pass a general exam (achieve good generalization).

Why Was the Adoption of Deep Learning Slow?

Let's now change gears a bit and discuss why it took so long for the ML field to adopt these "deep" methods. From our discussion so far it should be clear that the need for distributed representations came up very early in the history of ML, but the dominant thought around where deep methods should fit in was more on using them just for learning good representations or features of a problem to be later fed into some other learning system. In chapter 5, we mentioned why adoption was slow because of high experimental costs or infeasibilities. Here, let's talk about theoretical reasons why people were hesitant to explore the deep learning route.

Recall the 3 main theoretical aspects of the learning problem, namely, representation, optimization, and generalization. There were theoretical ideas in all those aspects that could discourage spending time and effort on deep methods:

- Representation — One can show that a shallow neural network can approximate any function if we allow the width to get as large as it wants to, which we have already stated. While that is true it is, it is also exponentially less efficient (per our discussion on depth). This is the fact that was somewhat ignored.

- Optimization — Depth will make the optimization problem "non-convex" which means there can be many locally optimum points one can reach and then stop before finding the global optimum (true minimum of the empirical risk). For that reason, most optimization theorists were urging practitioners to stick to convex optimization. It turns out that most of these optima arising from non-convexity in deep networks are saddle points (minima in one direction, maximum in another). Local minima happen to be quite rare, and once you find them they happen to be pretty good (not far from the global one). The geometry of these problems is still under active research full of interesting findings. What's known as mode-connectivity is one of such findings, describing the fact that in many of these deep network optimization problems, the minima are quite flat in some directions, meaning you can move around quite a lot in those directions without improving or worsening your results.

- Generalization — Recall that statistical learning theory used to strictly advise that one should limit the size or complexity of a model (relative to how many examples are available) i.e. do "capacity control", to be able to generalize well. So people knew they had to combine different learners and have them be diverse (per our discussion in the last section), but they did not want to put them all together inside a big model and to learn the "decomposition" inside that big model. Because that would increase the capacity and it was presumed that it would hurt generalization (due to overfitting). Instead, they tried to manually control how the learners are combined and kept the models separate such that the learning procedure in one learner would have no influence over the learning of another.

Not only have we conceptually argued that this would not work for decomposition learning, it has also been proven empirically to be a false assumption. Though, empirical proofs came only in the past decade! What we called "extended SLT" has only just begun to explain why (overparameterized) deep models don't overfit!

Research in Deep Learning

Deep learning has sucked in all the oxygen in the world of AI, and as such, the overwhelming majority of work in ML is now specifically in deep neural networks. Coming up with extended SLT theories that apply better to modern deep learning may be among the minorities of efforts in research. A much larger research effort, at least in terms of the number of researchers, is in extending the methodology and improving deep learning techniques. Yet, both these efforts are dwarfed compared to all the work that's being done to apply these techniques to more application domains and to overcome the challenges of real-world problems. We shall discuss the three thrusts of research separately as follows.

Extending the Theories

Algorithms that are useful are often too complex to analyze fully theoretically. Repeated non-linearities in deep networks are the most to blame as the most powerful mathematical tools we have are specific to linear operators. So, not surprisingly, our current theoretical understanding of why, how, and when a deep network (in particular, a deep neural network trained end-to-end via stochastic gradient descent) works, is not much further than what we have already covered in the last few sections. There is a lot more we'd like to quantitatively understand about deep networks, not just their own behavior, but also their relation to other things we already understand better such as kernel machines.

In the absence of any good holistic theory, lots of tools have been developed to look inside these networks and their dynamics during training in order to understand how they effectively do what we have arranged for them to do. There are various channels and angles through which one can study deep networks. Given that "learning" in a neural network occurs by adjustments of its weights (strength of connections between neurons — simply parameters of the model), the most obvious study channel would be to look at the dynamics (change over time as measured by optimization steps) of these weights. And indeed many studies do directly look at the distribution of model weights, how they evolve over the training process, and what characteristics are indicative of good generalization. However, the weights are not independent, they collectively achieve a representation of learning for generalization.

Other channels of study could look into more informative components behind the weights. Such as in some transformation of the weight matrices to get some underlying ingredients of them, or special combinations of them if you will. Spectral decomposition of the weights is such a canonical channel and indeed is used in active research.

Different sets of weights and parameters of the model give different values of statistical risk. Statistical risk is made up of the loss function. So the geometry of the loss function, the so-called "loss landscape", and how the change of weights move the network around in this landscape is another major channel of study.

After training is over, we have settled in a specific point in the parameters space, and the weights can no longer change. This fixed set of weights now represents the information about the training examples and contains it as much as the training process allowed it to. But the activations in the network keep changing with every new input. The activations all together change the input into something else layer by layer, and it is this process that is supposed to admit generalization. So we can study the behavior of the network, at this stage, beyond what the training set allows, using different distributions of the inputs and outputs. This is yet another promising channel of investigation.

Great, but what do we exactly want to get out of such studies? Again, it's first and foremost about generalization. Specifically, we want to understand better how exactly SGD in overparameterized networks with no explicit regularization works (i.e. doesn't overfit). We want to understand how well it works when it does. What kind of points in the loss landscape, in terms of risk minimization, it reaches (relation between their geometric properties and generalization)? How does it yield good representations, and where exactly, inside the network, is the information behind that representation sitting? And can we express, with more compact mathematics, what these networks do? Specifically, what are the ultimate kernels they are learning? And that could also help us understand kernel machines and their limitations better.

So what mathematical tools should we use to answer these questions and extend the theory? In fact, a variety of mathematical and mathematical-physics tools are being used. Let's emphasize two of them which are particularly more illuminating. They are the tools and techniques of 1) statistical physics, and 2) information theory.

Statistical physics is vast, to say the least. It does include dealing with situations where you have many particles and you cannot trace/follow the behavior of every single one. Yet, we may only need to pay attention to what they do collectively i.e. only their "distribution" holds relevant information for the macroscopic properties (what we care about). That is the same exact situation as in large neural networks. The concepts of particles and space (that the particles live in) are both abstract quantities, regardless of how much we think we are familiar with them because of our human experience. So the particles of neural networks could be individual neurons, their activations, their weights, etc., their space could be made of the set of possible values or attributes they could take, and a discrete time could be simulated by an optimization step or something similar.

With the oldest and most conventional statistical mechanics tools, it is easier to study larger networks and extreme cases such as the limit of infinite width or depth with neural networks. Obviously large or infinite size is relative. Here they are roughly relative to the number of examples available. Such studies were started back in the early days of ML, the 1990s, almost all the straightforward applications to the study of shallow networks came out then. In the past couple of years, post empirical evidence of the success of deep networks, many have been working on extending the work of those early days to *deep* networks. These studies can teach us a lot about the various phenomena that occur within deep networks. In the future, for more realistic (smaller) networks and more detailed theories of them, perhaps many lessons and tools of "mesoscopic" physics could also be utilized.

The other promising and powerful tool of study is information theory. In order to generalize, deep networks must hold exactly the right pieces of information from the input. That is, to keep what's useful for a "good representation" as well as the prediction task (output), and then throw out the rest. With information theory, we can track changes of information as it goes through the layers, track what gets kept in the weights, and trace the information dynamics during training. We can ask about the "distributional" properties of the behavior of any piece of the network rather than its "point-wise" behavior or property (such as input-output activations or correlations) upon a single input-output pass, or a single optimization step, etc. Before getting more into the weeds with what information theory can tell us, or has told us already about deep networks, let's mention the main context of research findings for deep networks.

We mentioned that the main questions are still around deep networks trained end-to-end with stochastic gradient descent (SGD). Many investigations of SGD, with mathematical tools similar to those discussed above, point at various aspects of its fascinating effectiveness. It has been observed for years now, in a variety of tasks, that SGD implicitly performs some kind of regularizations (essential to achieving good generalization). The success of SGD used to seem serendipitous. The reason is that, initially, it was being used not because of any theoretical understanding, but because it was the most feasible choice, and much faster than other methods for training neural networks. In SGD, during every step of learning, we use only part of the knowledge available to us, so learning becomes noisy but eventually, we can reach a quality that is good enough, with the upside that every step of learning is now much faster. So it seems like SGD is just a noisy version of regular gradient-based optimization, i.e. a noisy gradient descent (GD). It turns out that SGD is much more than noisy GD!

SGD seems to exhibit an implicit inductive bias, allowing the network achieve many interesting things that are under active research. Among those, we already mentioned regularization (a key to generalization). Other effects include settling on invariant and good features (properties that make yield more generality in the sense of making it more likely for the features to be useful for other tasks, and domains of the data), compression (in the sense of keeping minimally sufficient information for the task at hand), speed of convergence (doing it masterfully and conveniently, in such a way that is unexpected from just a noisy GD). Research work is underway to better understand how these things are all related.

We know for good generalization the network should only keep the relevant information for the task. And irrelevance has something to do with being too specific to individual examples. Information theory has contributed a lot to the quantitative study of such phenomena. Many are using information-theoretic tools to study where the relevant information, that controls generalization, resides within the network, how it changes, what happens to other pieces of information (non-so-relevant ones), and how it can be controlled.

For instance, controlled experiments can show transitions between underfitting phases and overfitting phases of a trained network if the information kept in the weights of the network are tuned above or below the "best" amount.[109] Other highly-illuminating results are in the study of information flow in deep networks using what's known as the information bottleneck method (IBT). Since we will have to talk about this method in volume II of the book, it's worthwhile to introduce it here.

As the name suggests, in this method the information has to go through a bottleneck and get distilled. Think: you have some information "on the left" to use and predict some information "on the right" (left and right are figurative here). You put a bottleneck in the middle to force only the most relevant information from the left to be stored, and be used to predict the right. Ravid Shwartz-Ziv, and Naftali Tishby (the original creator of IBT) most appropriately used IBT to study

---

[109] A. Achille, and S. Soatto. "Emergence of invariance and disentanglement in deep representations." The Journal of Machine Learning Research 19.1 (2018).

information flow layer by layer, through a deep network, and found interesting results.[110] They provided evidence that training the deep network with SGD first causes it to memorize all the examples[111], and then to forget the details that are too specific to the training samples which may hurt generalization. This can be thought of as a dynamic phase transition (a phase transition driven by time or training steps in this context).[112] Interestingly enough, this transition seems to happen exactly when the compression (or forgetting) starts.

This compression requires a lot more attention to be paid to as various studies seem to be seeing different signatures and consequences of it. In many studies, the geometry of the optimal points (the minima of empirical risk) in overparameterized networks have been found to be "flat" (a region with almost zero curvature or second derivative), and these minima are associated with low information stored in the weights.[113] Also if we look from a functional perspective (not from the space of parameters, but the functions they represent) the resulting functions (from the training) feature some low "functional norm". That means training overparameterized networks with SGD can lower the variation of the approximating function (the learned function) as much as possible. In other words, we find a solution that interpolates the data points, but otherwise maximally smooth, with much lower complexity than the model has the capacity to express.[114]

SGD is breaking many old assumptions. Recall our discussion of structural risk minimization, well, somehow SGD is taking care of it implicitly. Regularization in deep learning is still an open

---

[110] R. Shwartz-Ziv and N. Tishby. "Opening the black box of deep neural networks via information." arXiv preprint arXiv:1703.00810 (2017).

[111] Large enough networks certainly have the capacity to memorize all the examples with all their details. However, that doesn't mean any given example can be trivially (as in a database) retrieved. For actual retrieval and usage as memory see: Radhakrishnan, Belkin, and Uhler. "Overparameterized neural networks can implement associative memory." arXiv preprint arXiv:1909.12362 (2019).

[112] By analogy to physics, they argue that this is a transition from a drift phase (quickly getting to be around a good neighborhood — low empirical error) to a diffusion phase (moving around, back and forth, until settling on an optimal point by adding noise and forgetting the information that doesn't help). Such diffusion is describable by the relevant Fokker-Planck equations here (i.e. dynamic of the probability density of your variables, in this case the distribution of the weights of the network).

[113] That is low Fisher information which involves the second derivative, relating it to curvature in the parameter space. See A. Achille, G. Paolini, S. Soatto "Where is the information in a deep neural network?" arXiv:1905.12213v5, (2020).

[114] C. Ma and L. Wu. "On the generalization properties of minimum-norm solutions for over-parameterized neural network models." arXiv preprint arXiv:1912.06987 (2019).

theoretical problem.[115] There are many explicit methods for regularizing neural networks but what seems to be dominating successful practices in deep learning, is just SGD often with "early stopping"— cutting out the optimization before it's fully exhausted. SGD also happens to be much faster than expected. Almost equivalent to full gradient descent (GD) in terms of the number of steps it takes to reach convergence or some good enough neighborhood.[116] But like we mentioned earlier, it's much cheaper compared to GD, per step. This fact is most likely linked to those memorization and forgetting phases of training we just discussed.[117]

To better understand these phenomena, how they are all related, and can be controlled, many are studying the "loss landscape" and the "trajectory" of optimization (in the space where the parameters live) with more sophisticated tools borrowed from mathematical physics. Meanwhile, such efforts and alike, at least partially motivated by the empirical successes of SGD in deep learning (a highly non-convex optimization problem), are teaching us a lot of new things about optimization in general. In that sense, it's much harder now to separate the optimization and generalization aspects of learning from each other. It is inevitable for the representation aspect of learning to eventually join the party too, once our understanding of generalization gets deeper.

Putting SGD and the details of optimization aside, we need to better understand how deep networks and their generalization relate to kernel machines. That is because we think we understand kernels the most, at least within the statistical learning theory (SLT) of the 90s. Their generalization properties are much more explicitly controllable using structural risk minimization and capacity control — see "statistical learning theory" section. However, as we discussed in the deep learning section, deep networks, or at least parts of them, are implicitly learning some kernel within their overall learning. This fact alone suggests that we don't have the full picture on the generalization theory of kernel machines either. Going even further, one could argue that kernels are a subset of neural networks.[118]

---

[115] Though, many advances are under way, for instance, see: S. S. Du, W. Hu, and J. D. Lee. "Algorithmic regularization in learning deep homogeneous models: Layers are automatically balanced." Advances in Neural Information Processing Systems. 2018.

[116] S. Ma, R. Bassily, and M. Belkin. "The power of interpolation: Understanding the effectiveness of SGD in modern over-parameterized learning." International Conference on Machine Learning. PMLR, 2018.

[117] Ibid.

[118] Z. Allen-Zhu and Y. Li. "What Can ResNet Learn Efficiently, Going Beyond Kernels?." Advances in Neural Information Processing Systems. 2019.

In some cases, the kernels that are represented by deep networks can be stated explicitly and compactly, in which case they can be of great utility to stage further theoretical investigation. Neural tangent kernels (NTK) are a great example here.[119] NTKs allow for studying the function that is represented by the whole network directly, rather than indirectly through its parameters (the weights). It is shown that in the limit of infinitely-wide networks, NTK is only a function of the architecture of the network, meaning the training process doesn't affect learning once the architecture of the network is fixed. NTK can also suggest explanations for why early-stopping (part of the implicit regularization we mentioned earlier) is a good practice and when so.[120]

Ultimately, all these various angles and tools of study will have to converge to a self-consistent and holistic picture. Different insights should point us towards some unified theory here. That is, we should be able to get some unified and comprehensive answer from the theory if we ask what happens exactly if we do X or Y to the network, or use a different objective function, etc. It's unclear whether such a theory exists, but if it does, we are currently pretty far from it.

Extending the Methods

Think of deep learning so far as a successful "proof of concept" (mainly in the case of supervised learning with successful applications in various domains, from research to being deployed in the commercial world). That is, a proof of concept on this method of learning. Now it needs a lot more work to push it to various cases, beyond those of the proof of concept, and to make it truly robust. Recall our discussion on learning types, or see section: types of learning. We need to bring deep methods into all those different types. Below is an unordered and non-comprehensive list of big-theme topics that can nevertheless take you for a short, but deep, trip into the world of deep learning research on extending the method:

- **Explainability**: Combining learners leads to a lack of straightforward interpretability, that individual learners may possess, leading to a black-box (combined) model. Deep learning by definition fits in that class. Therefore, providing an explanation as to why the network chooses some

---

[119] As the name neural tangent (the partial derivative of the function represented by the neural network with respect to its parameters) implies, the kernel is constructed by evaluating the neural tangent at two different (input) points and taking the expectation (average) over different parameters of the network. See: A. Jacot, F. Gabriel, and C. Hongler. "Neural tangent kernel: Convergence and generalization in neural networks." Advances in neural information processing systems. 2018.

[120] Ibid.

output over another, in a way that can be helpful for a human decision-maker, is no trivial matter. In the absence of any "natural/default" way to interpret the choices of the network, a secondary process must be created, and likely be learned (as it is context- and task-dependent) to generate a useful explanation.[121] It's important to note that we humans do exactly the same thing when asked to explain ourselves. For instance, in my introspection, I am aware that my thoughts and any explanation of them are always totally separate thought processes. It's as if one person is explaining another person's mind.

- **Apply to other data structures**: Traditionally, research has been conducted with the kinds of data that can naturally be embedded in a (high-dimensional) Euclidean space (the space we're most familiar with and know how to measure the distances within, you know, the space of everyday life locations). That includes data types with a list of features or a grid of features, like a sequence, or a 2D image where pixels form a 2D grid (a matrix with every cell representing a value of a feature), or more generally, a higher dimensional grid (list of list of lists, and so on). In all these cases we can think of the raw input as some long vector, albeit, in most cases, there are many different ways to build up such a vector.

Most often, it shouldn't matter if your learning method (including the architecture) is robust enough. But there are many interesting problems where the raw features don't naturally conform to a grid. That is, they have some special geometry, and the geometry of the collection of features is an important and continuous feature that cannot be trivially vectorized. Think of the boundary of a 3D object, i.e. a 2 dimensional curved surface in the 3D space. This is an example of what mathematicians formulate as a *manifold*, although the loose usage of this word in ML drives real mathematicians crazy! Manifold learning constitutes an important area of research.

Then you have other mathematical objects like *graphs* that don't even have a geometry (unless they are embedded in some space, which is no trivial matter) but have other properties to be accounted for. *Graph neural networks* try to address that. A very active field and has been for a few years now, growing year over year. It's important to note that graphs are dizzyingly prevalent in the real world[122], yet they only make up for a tiny fraction of commercially-deployed "AI". The

---

[121] For a status review of such work see: N. Xie, et al. "Explainable deep learning: A field guide for the uninitiated." arXiv preprint arXiv:2004.14545 (2020).

[122] Simply because graphs are very general in that they can represent almost any other data type. Though, there may be many ways to come to such representations, and how useful an equivalent representation in graphs can be, is a separate matter.

applications of graph neural networks in medicine and biology are enormous, standing to change the way research pipelines are set up by adding abundant "AI"-computational assistance especially in the form of good hypothesis generation.

- **Integration with broad prior knowledge**: State-of-the-art networks for image classification are convolutional neural networks (CNNs), which use something called a convolution and you can think of it as putting some (very broadly applicable) prior knowledge into the network. The prior knowledge used there is "translational invariance". That is, for the task of recognizing an object in the image, it doesn't matter where the object appears in the image i.e. it is translationally invariant. This is an instance of broadly applicable knowledge of our everyday human experience with fundamental roots in physics. In CNNs, this is put in by hand, by design. It is OK to put this by hand, because again it is very broadly applicable, with no extra human labor overhead. There is another broad prior knowledge that is designed into CNNs as well, and that is *scale invariance*. The size of the object, how big or small it appears in the image, also shouldn't matter for the task of recognizing it. That's put in by hand through "pooling" layers in CNNs.

Similarly, there are other problems in which other priors need to be designed into the network. For instance, the priors that can and should be used for learning graphs, or manifolds we just mentioned, are way beyond the simple convolutions or pooling used in CNNs.[123] This too is a very active and exciting area of research, being pushed by many with strong backgrounds in physics and mathematics. Sometimes this field is referred to as "geometric deep learning".[124]

- **Augmentation with (differentiable) memory**: So far in the "successful proof of concept" we have a promising technique to approximate any function in principle, using many examples of input-output pairs. Now, that sounds powerful, yet asking any (conventional) deep network: "what example did I just show you?", or asking it to copy the example out exactly as it received, falls outside its scope. Retrieving from memory, or copying to memory, and retrieving the copy, is an elementary capability of any modern computer (Von-Neumann representation of a Turing machine). That's because Von Neumann's architecture comes with both a memory and a controller. The controller can read from memory, turn it into something else, and write it back to

---

[123] T. Cohen, M. Welling. "Learning the irreducible representations of commutative lie groups." International Conference on Machine Learning. 2014.

[124] Bronstein, Michael M., et al. "Geometric deep learning: going beyond Euclidean data." IEEE Signal Processing Magazine 34.4 (2017): 18-42.

the memory unit. The "turning it into something" can in principle be taken care of by some neural network, because that's what functions do after all. So it sounds like all that is missing to make a neural network become a neural *computer* (capable of implementing algorithms) is to augment it with an explicit memory unit. This idea of a "Neural [network based] Turing Machine" was pursued explicitly by Alex Graves et al. in 2014.[125] And has subsequently inspired lots of activity in this very important direction to date.

Let me give you an idea of how this works in its simplest form. One can store the activity of say a layer of the network on a memory cell (a single cell can contain a vector of the neural activities). The question is where should you write it, in which memory cell? Well, we don't know, so we could write it in every cell, but as much as that memory cell already contains something similar to the vector we want to write in. If the similarity (overlap of the vectors) between the vector already in the cell and the to-be-memorized vector is zero, nothing will be written in that cell. You can also read out from memory in the same way, you come in with a vector containing some information about what you want to read out. Every cell may have something to contribute, and we can read them out as much as they are similar to the read-requesting vector. You can think of this as a *soft* read-and-write, as opposed to a *hard* read-and-write in conventional computers, where every cell is assumed to have something absolutely unique, hence it must also have an absolutely unique name or address. In hard read and write, there is not much you can do with that if you don't know the exact name of the cell you want to reach. Soft *addressing* means working with smooth functions which we can parametrize, differentiate, and use differentiation to make adjustments upon seeing examples i.e. learn from examples, learn many things including how to address and what to write in, when to write, and so on, in ways that are not possible in hard addressing.

It's important to note that lots of inspiration and renewed interest in neural networks in the 80s also came from the idea of representing memory by a network that learns softly, which started with the pioneering work of Hopfield on associative memories, aka. Hopfield networks. That also gave rise to ideas for *content-addressable* memories which in this sense can be thought of as the parent category of these differentiable memory architectures.

• **Data efficiency using transfer learning:** Using prior knowledge and augmentation with explicit memory both can help to solve a learning problem using fewer examples i.e. with more

---

[125] Graves, Alex, Greg Wayne, and Ivo Danihelka. "Neural Turing machines." arXiv preprint arXiv:1410.5401 (2014).

data efficiency. We can also use the 'implicit' memory possessed by the learned network (implicit in the weights of the network) to learn other tasks with more data efficiency. This is the core idea of transfer learning, and when successful, can result in very high data efficiency.

Transfer learning is a broad topic with many variations depending on the purpose and setup of the transfer. For instance, multi-task learning and domain adaptation could both be considered special cases of transfer learning. In multi-task learning, an intermediate shared representation is learned and simultaneously used for multiple tasks performed by one model performing them. In domain adaptation, the model is transferred to be used for the same task but in a slightly different context which means the distribution of the input examples have changed (e.g. you are classifying fruits based on their taste, and in a new climate the apple is more bitter than in the climate you initially learned the taste of fruits at).

- **Meta-learning:** During the optimization process of learning, we are incrementally adjusting the parameters that make up the function we are approximating. But there are many other parameters that are not modified in this learning process, such as those that specify the representation of learning, and the optimization process, i.e. the learning process itself. These parameters that are chosen outside of the learning loop, are mostly referred to as hyper-parameters [126] Learning good values for hyper-parameters can be a secondary learning problem, an outer loop encompassing the inner loop representing the main learning task. This kind of learning is referred to as meta-learning and sometimes goes with the grandiose title of "learning how to learn".[127]

But what does *"good* values" mean? Meta-learning can be used for hyper-parameter optimization to improve the generalization performance of the learning problem at hand. In this case, good values for hyperparameters are typically those that result in some optimal model capacity resulting in the best generalization performance. But meta-learning can be a lot more than that, namely, it could be much closer to "learning how to learn". That is to learn the approximately best

---

[126] Hyper-parameters often refer to those "outer-loop" parameters that are explicitly tunable such as the learning rate, choices of activation functions, or number of hidden units in a layer, etc. In this discussion we are using a broader definition for hyperparameters, covering all choices in a learning system that are not adjusted by the learning process including parameter initialization, order of input examples (as in "learning curriculum"), and so on.

[127] S. Thrun and L. Pratt, "Learning To Learn: Introduction And Overview," in Learning To Learn, 1998.

setup of the learning algorithm for some other set of tasks. Think figuring out how best to initialize the weights (of the network) at the start of learning, or finding the best network architecture (known as Neural Architecture Search), not just for the given task but perhaps a collection of related, or similar tasks, or figuring out the best choices (of hyper-parameters) in a slightly different context for the task (known as domain adaptation/generalization). Other examples include how and when to ask for new labeled examples (the case of *active* learning); how to order tasks when learning multiple tasks with one model (the case of *multi-task* learning), and in what order is it best to feed the examples into the learning algorithms (the case of *curriculum* learning).

Given that, 1) inside meta-learning, there is always some learning involved, 2) the purpose is to be transferring what we learn about the hyper-parameters, or other learning choices, to another situation or problem, meta-learning can be said to be intimately related to transfer learning. However, they are distinct, in the sense that transfer learning is more explicit, it wants to recycle what it learns about the parameters for new or downstream tasks directly, whereas meta-learning focuses on transferring what it learns about the algorithm of learning itself.

Many see meta-learning as the main candidate for the next revolution in ML. Think in the sense that, just as deep learning automated feature engineering, meta-learning could automate algorithm engineering. Although it is a premature and vague statement, it does highlight the big potential of meta-learning, and the need to establish it better.

- **Learning better representations that separate out causes of variation in the data:** Representation learning is a huge field with no crisp definition, but qualitatively the goal is very clear, and it's about finding representations that "make life easy" no matter what you choose to do with it. Transfer learning mentioned above can be considered an instance of representation learning. The better the learned representation, the more successful and easy will be the transfer of it.

One goal within representation learning is to reach a set of features such that the prediction task or function approximation task reduces to finding the right linear combination of those features. Such features in the context of representation learning are called "causal factors" as they are the causes of variation in the raw data we see. Note that this is different from (and not to be confused with) working explicitly to discover the cause of an event, or the root cause for something, which is the subject of "causal machine learning".

Unsupervised learning and semi-supervised learning (when some examples are labeled and others aren't) have lots of overlaps with representation learning, but they are all distinct, with unsupervised learning being the widest and most open field of all. Representation learning has many overlaps with reinforcement learning (RL) and control problems too. Representing the states of the agent and the environment in a rich yet minimal (representing only what's relevant for good decision-making or acting) fashion, can significantly boost RL and control methods.

- **Learning probabilities and generative modeling:** This one is huge because it is a basis for unsupervised learning. So let's take a step back. What we get out of a typical neural network (the "successful proof of concept") is an approximate value for the function we are approximating at a particular input value, i.e. a prediction (for the value of the output). And modern deep networks can be great function approximators, meaning their predictions can be highly accurate. However, a quantified confidence for any given prediction is still missing. That is, we'd still want to know a probability value for predicted outputs. Sometimes (using a function called softmax) we can get prediction values that sum up to 1 and may look like probability values but there is no principled way to think of them as so. This is an inherent short-coming because knowing the probabilities can be very useful and sometimes necessary. It can be necessary for many decision making tasks such as those involving scarce resources. It can be very useful in unsupervised tasks, for instance, in generating a true sample that we haven't seen among the examples provided but we could have (known as *generative learning* or learning a generative model), or to predict missing values in the data. These were features that came out of the box with Bayesian networks, at the advent of modern AI (see section pre-modern to modern AI) but are now missing from simple deep learning.

We can borrow the idea of Bayesian networks and create structured probability models that are deep. That means they'd have many levels of latent variables (unobserved variables that can have explanatory information about the observed data). The model is often a joint probability of all the variables involved (observables and latent), and inference means inferring the probability of a particular set of variables (called *marginal* probability because it eliminates the rest of the variables which are not in that particular set). However, in this field of deep-probabilistic-models, inference often just refers to finding the marginal probability of latent variables given data. We can use deep learning techniques for inference in these models because the old inference techniques, used in Bayesian networks, are not practical in the case of dense networks with too many latent variables (much more than observable ones), as they would involve intractable computations. This is true whether we want to just compute the marginal probability of predictions (which involves computing or estimating the sum of probabilities for every possible outcome for predictions known as the partition function[128]), or inferring the latent variable probabilities. In both cases, a bunch of tricks that involve *sampling* techniques one way or another (using the neural network itself), along with clever objective functions, are used to approximate the probabilities of interest.

We can divide the tricks to overcome inference of probabilities into *direct* and *indirect* methods. Direct methods would try to estimate and approximate the true probabilities of interest. That includes the direct estimation of the partition function, estimating its gradient, learning it along with other model parameters using the contrast against some noise baseline (known as *noise-contrastive estimation "NCE"*), or bypassing any explicit calculation of the partition function. On the other hand, indirect methods of learning probabilities, would replace the true probability with a different probability model and try to make it closer to the true probability through an optimization process. This approach is widely known as approximate inference or variational inference. Approximate inference typically works by maximizing a lower bound (known as the evidence lower bound, or ELBO).

We can also look at the act of inference as applying a function that performs some algebra and computations on other functions that are learned by a deep model. So we should in principle be able to include this whole algebra/computation part (inference) as part of the total function learned by a (bigger and different) deep network. Recall how deep learning can learn the right decomposition, so as to spare us from a manual process. We can cleverly try to apply this to learn a specific inference task, thinking of it as a task that needs to be implicitly broken down into learning and inference in the learning process of a more sophisticated model that includes both. This the clever idea behind *learned approximate inference.* The most famous and widely used examples of learned approximate inference are variational autoencoders (VAEs). VAEs simply plug a generator network (decoder) into the end of an inference network (encoder), and the whole network is learned through one learning process. Having said that, they perform only a specific kind of inference, namely inferring the latent variable vector from the vector of the observed example. This model would know about the (approximate) probability values in the space of the training data, just enough to perform the particular inference task well.

---

[128] This sum is known as the partition function because it can be thought of as a function of partitions where every partition contains all the possible outputs that share the same likelihood probability. The biggest partition which contains most output cases contributes the most to the partition function. For physicists, output cases are actual system configurations or particular arrangement of particles, and partition function implicitly holds a rich knowledge of a statistical system, and as such shows up almost everywhere in dealing with probabilistic quantities.

In explicit (whether through direct or indirect approximations) models of probability, we always learn some probability. On the contrary, with implicit models we are not approximating any probability, we jump straight at the task that we would have used a probability model for, had we had it, such as sampling from it, or conditioning it on some context, etc.

In implicit models, it is recognized that **probabilities are just means to an end**! Such models bypass challenges of explicit probability calculations to generate things that require knowledge of probability. In them, we know implicitly what is going on in the probability landscape. For instance, we gain an implicit knowledge of the partition function, when explicit knowledge of it could require intractable computations. Therefore, all implicit models can be said to be learning probabilities indirectly. Generative adversarial networks (GANs) are among the most popular and successful examples here. The inference task in a GAN is taken care of in an entirely indirect and implicit way. There are various formulations of GANs. In the original setup, implicit inference happens through playing a game between a generator (generating a sample of input data) and a discriminator (judging how realistic or true the generated sample is), while both are trained together. The following matrix exemplifies our categorization:

Examples of probability estimations in DL	Explicit	Implicit
Direct	Partition function estimation	N/A
Indirect	VAEs	GANs

- **Robustness and stability**: Reproducibility in research used to be an issue in machine learning in general, but that has been mostly addressed in the past decade through common datasets and standardized tasks. However, the sensitivity of many methods we discussed to slight changes, or perturbation in the training procedure (think very modest changes to hyperparameters including choice of optimization algorithms), or in the data the trained systems are tested on, is still quite an open and underdeveloped field of research. This is not just a challenge in applying the method to real-world applications, it might turn out that the methods themselves would need to change to unrecognizable degrees to truly address robustness and stability issues. That's because the problem is often the method itself that creates vulnerabilities such that simple tweaks, or customizations for a particular domain of application, wouldn't resolve it.

Consider the case of adversarial examples, which are examples intentionally designed to fool a system into misclassifying the example. For instance, the image of a building could be modified with some noise barely detectable by humans, but sufficient to fool an image recognition system to falsely classify it as an ostrich instead.[129] Adversarial examples are typically just a linear noise in raw input space, overlaid on top of an input. Being fooled by such linear noise means the machine learning system is not truly non-linear in terms of the input, even though it may be highly non-linear in terms of its parameters as deep convolutional networks (used for image recognition) certainly are. A robust non-linear model of the input space should be able to ignore the noise, and not get fooled by irrelevant changes, i.e. be robust! Lack of robustness here can open the door to adversarial attacks, and pose (machine learning) security threats.

We ultimately need good theories to tell us exactly in what ways, and when, ML systems can break, or not perform sufficiently well. Given that theoretical research is still far behind modern practice. This research here still needs to be conducted mostly within ingenious trial and errors, possibly spinning into new directions full of new lessons and surprises.

- **Bayesian deep learning:** Even when things work, and when we can calculate probabilities to quantify uncertainty, there still remain uncertainties about the process of finding the probability. There can be an *epistemic* uncertainty that results from having little evidence, or from gaps in our knowledge. We can still quantify uncertainty about the events out there in the world given limited knowledge, but we still would have to quantify our uncertainty in doing so. In other words, predictions should come with error bars on them. Though the ultimate goal is really proper decision making, which one way or another, should be informed by uncertainties in the process of quantifying uncertainties.

Normally that's addressed within Bayesian learning, where we get the full posterior probability measure, and we use it to integrate over predictions so that all possible models/estimates contribute accordingly to their posterior weight (our degree of belief in the correctness of the model). The field of Bayesian deep learning tries to address the problem of quantifying confidence in predictions directly, by putting probability models on every parameter of the model and on the output of the

---

[129] Szegedy, C. et al. "Intriguing properties of neural networks", arXiv:1312.6199 (2013).

model, given those parameters and the input data. In this way, it can explicitly compute the uncertainty in the predictions of the network. This is crucial because our appetite for uncertainty varies depending on the situation, and such uncertainty quantification allows us to settle for an acceptable, but not perfect, quality of results when little amount of data is available to infer from.

- **Other:** There are many other research avenues that we don't need to cover, nor have the room for. An example would be causality, and working on models that learn, or know, about cause and effect relationships, as well as use them for understanding and reasoning. Yet, the obvious field we left out that may eventually encompass all the above is the vast field of reinforcement learning and embedded (acting in the environment/world) agents, which have been revitalized by deep learning methods, and are the hottest in terms of the attention they receive among all the other research areas we covered above, and rightfully so, for the great potential they bear.[130] A huge challenge there is in acting on long time scales. We humans excel at achieving high-level goals by breaking things down in hierarchies that become natural to us but are highly non-trivial for machines. We need machines to discover natural hierarchical decompositions too, on their own. As we mentioned, deep learning is an instance of decomposition learning, and in that sense, there is great hope for deep methods to help machines break a high level task into several levels of abstractions each associated with a different time scale. Another challenge is in learning the objective functions, or parts of them, and not relying entirely on a designer of the machine learning system to specify them. That's crucial given that often what we precisely want from a machine is within our tacit knowledge, and by definition, we aren't that good at specifying them with sufficient explicit details. Techniques such as inverse reinforcement learning, or imitation learning, take a step in this general direction although they only begin to scratch the surface on what's really needed for autonomous machines. That brings us right to the challenges of applications of the ML methods in the real world.

Overcoming Application Challenges

Overcoming application challenges qualifies as a separate hardcore research track, not just because the real-world imposes its own unique challenges, but also because our theoretical understanding of deep learning practices isn't sufficient for the application part of the methods to be straightforward

---

[130] For an extensive overview, see Y. Li, "DEEP REINFORCEMENT LEARNING", arXiv preprint arXiv:1810.06339 (2018).

engineering tasks. Most novel applications become their own research projects, except not necessarily in an academic environment, and often accompanied with false business promises.

Let me mention a few topics on the research side of overcoming real-world challenges with broad relevance beyond any specific application or domain:

- Computational frameworks to make both training and inference faster, and to a lesser degree make development, testing, and scaling of big systems easier.
- Special-purpose, or accelerated hardware, designed specifically for certain ML workloads.
- Building large real-world datasets, either privately or publicly.
- Data augmentation techniques to learn well from less data.
- AutoML — automating many parts of creating a successful ML model.[131]
- Model compression techniques to cope with memory-bound environments, such as in mobile devices.
- Model pruning to cope with compute-bound environments such as in edge computing.
- Model distribution: the goal is to properly distribute the learning and training data over multiple machines.
- Federated learning: a form of distributed learning where the aim is to not move any data out of the local node where it belongs, to address data privacy, data security, data access-right issues, and more. The idea is to move the learned parameters around instead.
- Privacy-preserving learning: Omitting sensitive fields from the raw data is still far from truly protecting privacy. Privacy-preserving techniques aim to not only hide sensitive data or meta-data, but also not allow for the possibility of them being found through reverse-engineering or effective inference, which can be a safety issue.
- It is not even sufficient to simply hide sensitive information, and not use it for training, since what is used for training may already allow the network to have implicit knowledge of what should have been kept private, and remains exposed to wrongful extraction.
- Ethics issues and bias: We are ending with this, as it is among the most critical challenges. Deployed ML systems in the real-world make predictions and influence decisions, which influence real lives. From the datasets, which are used for training the models, to what the

---

[131] The principled way to do this is through meta-learning, but most often many heuristics are used until further maturation of meta-learning techniques.

models are for, relative to the data they use, there can be many blind spots with ethical implications to pay attention to. This is yet another open and underdeveloped field of research.

The list of research fields goes on and on. As we talked about, there is an unprecedented amount of capital being spent globally on ML research, and thus, a lot is happening in the world of ML/DL research, which no single book alone has enough room to cover all. The goal here was just to give a sense of the comprehensive research box, and its boundaries, so that the reader can start asking questions that fall outside the box!

Chapter 7

AI as Hybrid Systems Beyond Deep Learning

Limits of Deep Learning

We just reviewed growth opportunities for deep learning (DL) that are being investigated in research. Now, we'd like to ask what are the limitations with DL practice that falls outside all these research areas or at least outside what's currently being done in these areas. In other words, how much more mileage can we get out of DL before we hit dead-ends? Or is it that there aren't going to be any dead-ends and deep learning is going to become the master algorithm? Do these questions make sense? How so and how not so?

Let's start with current limitations. People describe deep learning as AI, so it's fair to put a high bar for it to pass. Therefore, anything that deep learning can't yet do when other methods (e.g. in classical AI) or humans can, can potentially be considered a limitation with the method. But we don't have to go that far. Practically all those research areas, we considered in chapter 6, are exposing limitations with deep learning. And, of course, there were many that we didn't get to cover. Some of these problems are considered harder and longer-standing challenges than others.

An example of such long-standing problems which we didn't cover, is the problem of "catastrophic forgetting" in neural networks. The concept is ported over from cognitive science and it refers to the phenomenon where new learning overrides previous learning. In plain deep learning, new training for a new task changes the weights in the network, and therefore, the ability to perform a previously learned task is lost. The research to overcome catastrophic-forgetting is ongoing with interesting avenues and preliminary results to explore.[132] However, it still remains a big enough of a challenge to justify calling it a limitation.

But is it fair to call a limitation with current DL methods, a limitation with DL in general often phrased as "DL can't do X"? What if, later on, a simple extension (such as a new component in a dynamic architecture), integration with a broad prior, a clever meta-learning policy, or whatever you consider a promising extension, overcomes something that we currently call out as a big limitation? It's highly unlikely that any of these alone can solve big limitations, such as the "catastrophic forgetting" problem. But that's not the issue here, the debates are more on what we should consider deep learning going forward, and whether it matters or not for research efforts, (scientific) communications, fundings, etc. This is a serious issue since many folks nowadays, in both private and public institutions, have serious trouble getting funded if what they do is not called AI, and it almost certainly won't be called AI if it's not considered to involve deep learning.

In early 2018, Yan LeCun one of the pioneers in deep learning and head of Facebook AI research wrote:

> "OK, Deep Learning has outlived its usefulness as a buzz phrase. Deep Learning est mort. Vive Differentiable Programming! Yeah, Differentiable Programming is little more than a rebranding of the modern collection Deep Learning techniques, the same way Deep Learning was a rebranding of the modern incarnations of neural nets with more than two layers. But the important point is that people are now building a new kind of software by assembling networks of parameterized functional blocks and by training them from examples using some form of gradient-based optimization."[133]

Later by the end of 2019, he writes:

> "Some folks still seem confused about what deep learning is. Here is a definition: DL is constructing networks of parameterized functional modules & training them from examples using gradient-based optimization. That's it." [134]

This is more of an explicit description of the programming logic than anything else. As a definition, it does not say much since it is a very wide and barely-constraining statement. For instance, the

---

[132] G. Parisi et al. "Continual lifelong learning with neural networks: A review", Neural Networks, Volume 113 (2019).

[133] https://www.facebook.com/722677142/posts/10156463919392143/ —last accessed Dec 2020.

[134] Ibid.

"depth" requirement is only implied here. The constraining part is the choice of "gradient-based optimization", and as such, the name differential programming is perhaps appropriate. It's a large umbrella for all sorts of activities inspired by DL and can be considered to be in the same vein. At any rate, the message we are getting is that the DL community would like to keep the term DL as the overall label for what they do and thus it will not have a fixed definition, i.e. subject to change over time. This sentiment was further echoed by Yoshua Bengio, a deep learning pioneer who leads the largest deep learning research group outside the industry, based in Montreal, Canada. He writes by the end of 2019:

> "Deep learning is inspired by neural networks of the brain to build learning machines which discover rich and useful internal representations computed as a composition of learning features and functions. Note how this definition is a goal and does not say much about HOW we achieve that. ...because it's an open problem how to achieve that goal best, the term deep learning is indeed aspirational like AI or machine learning." [135]

Therefore according to the biggest names in DL, there is no definition for DL, rather it's a placeholder for whatever may come. This sounds totally fine and reasonable, except that obfuscating-marketing and ill-messaging practices by a few other parties, such as some giant tech companies have caused troubles for those who aren't exactly "deep learners" (what deep learning researchers like to call themselves). That means that there could be uses for good definitions. It was in this spirit that we also tried to give our own definition in chapter 6, though more for pedagogical purposes (see section "deep learning as a case of decomposition learning").

All in all, the established facts are that 1) there exist varying opinions about what DL is and should stand for. We'll come back later to what we should make of that if any. 2) There exists somewhat of a consensus that many of the current limitations are not going to get solved by mere "more-of-the-same work". They likely require some new or perhaps outside-the-box thinking. What are some of such challenges that are front and center in the slightly longer-term research agenda for deep learning?

Current DL models lack context and background knowledge about how the world works. Creating better "world-models" that can provide sufficient background knowledge is part of current efforts to address this challenge. However, this may be best approached by working on

---

[135] https://github.com/MontrealAI/MontrealAI.github.io/blob/master/aidebate/yoshua definitiondeeplearning.png —last accessed Dec 2020.

embodied intelligence, as a better understanding of the environment could be key for fulfilling the goals of the agent. The agent perspective can, in turn, open the door to addressing other challenges such as a more inherent understanding of causality.

Another setting for helpful research in this direction is provided by "grounded language learning", where the goal is to not just be able to perform the natural language processing of some piece of text, the model should learn what words refer to what objects, agents, or relations in the environment, and be able to do visual reasoning about it and articulate that reasoning in natural language as well. As you'd expect, this theme is considered a promising avenue in the field of natural language understanding.

Another area of challenge is to create models that are capable of stronger generalizations. That means becoming applicable to not just the instances from the distribution of the training data, but also to other data distributions and even from different domains. This is the so-called "out-of-distribution generalization". A great example of models capable of out-of-distribution-generalization is logical rules in GOFAI systems. This brings us to the heavy-weight challenge for DL, namely general reasoning, some systematic way to integrate with prior knowledge, and flexibly composing multiple high-level concepts or representations for *systematic generalizations*, again as was the case for classical AI and in rules of logic.

Needless to say, DL researchers are working on many forefronts of current limitations. The above-mentioned are just some of the broader challenges sitting between current methods and any "true AI", "general intelligence", "master algorithm", "human-like intelligence", or pick your favorite ill-defined or definition-less term here.

While many DL researchers seem confident in the pace of progress towards overcoming these challenges, other AI researchers (from outside the deep learning research) take the current limitations to be more serious, sometimes taking the positions that (at least current) deep learning can never lead to "true or general intelligence" without significant rethinking or extensions. Such claims typically come from two different groups: 1) from more traditional AI researchers, and 2) from cognitive scientists. The argument of the first group often is that rules of logic and symbolic AI are quite general and intelligent systems must one way or another go through them. For instance, as we reviewed, first-order logic is designed to be the most general language to describe objects and relations since we mostly perceive our reality in that way. So it's most likely needed to organize the AI system's knowledge in a similar fashion to best mirror the world we perceive and it perceives! Then there is the argument by cognitive scientists that humans perform all the tasks (that DL has been successful at) very differently.[136] A well-cited narrative advocates for a research direction towards:

> "Seeing objects and agents rather than features, building causal models and not just recognizing patterns, recombining representations without needing to retrain, and learning-to-learn rather than starting from scratch." [137]

DARPA, whose support and initiatives have been behind major research breakthroughs in AI over the past couple of decades, calls current deep learning part of the second wave of AI technology, i.e. the "statistical learning" (the first wave being handcrafted knowledge and feature engineering). Considering a number of challenges with deep learning, DARPA proposes the next (future) wave to be of "Contextual Adaptation", where "systems construct contextual explanatory models for classes of real-world phenomena". The abstraction and reasoning abilities of third wave systems are supposed to be far superior to current DL systems.

Now let's revisit the question of: are any of the challenges we mentioned truly limiting the future of DL? What future work should we still call DL? Well, from our discussions it should be clear that depth is a necessary feature no matter what you do. So it doesn't quite make sense to question the future of depth. But depth is just a feature simply reflecting our world, not a system, method, or algorithm. Differentiability in vector spaces and end-to-end gradient methods can on the other hand be questioned. Many in the DL community acknowledge that and are working to extend deep learning to where it can also perform what symbolic reasoning systems easily can. Yoshua Bengio has promoted for DL research to go beyond subconscious perceptions and intuitions (called system 1, the fast system), and reach conscious thoughts and reasoning (system 2, the slow system)[138].

For those who still believe end-to-end differentiability can take us "to the end" in AI without symbolic methods interweaved with it somehow, let's point out the case of "non-analytic phenomena" that so far I haven't seen discussed anywhere in the vast AI literature. Roughly speaking, non-analyticity is about non-differentiability at certain points or regions in the domain of

---

[136] G. Marcus "Deep Learning: A Critical Appraisal." ArXiv, (2018).

[137] Lake, et al. "Building machines that learn and think like people." Behavioral and brain sciences 40 (2017).

[138] In his book "Thinking fast and slow", Daniel Khaneman identifies these two systems in the brain as being qualitatively very different, and the appropriate choice of name, 1 and 2 is to prevent association with any particular brain region or process.

a function. In theoretical physics, these points are of high interest because they can signal qualitatively new phenomena, and hence we call them non-analytic phenomena. Non-linearities in deep learning cannot help spit-out the non-analytic points of interest in a non-analytic target function.[139] "Differentiable programming" can approximate some function to some accuracy given enough examples of input-output pairs to compute effective gradients.[140] But in mathematical theories, especially concerning non-analytic phenomena, one can never do that. That is exactly the business, when you cannot do the experiment, at least not cheaply or in a timely manner, and doing the regular exact math is also not an option. This argument is sufficient to rule out differentiable programming ever being able to "do science" by itself, although it can help scientists in their data processing or even do regular math, and it already has.

One has to be careful here with any statement of the sort "oh Neural-Networks or DL can't do X or Y". This is in fact the wrong way to talk about things and it seems like moving the goalpost for deep learning. History is full of such statements that were proven wrong later. Such statements were not on any rational basis to begin with, most often based on false guesses on the difficulty of some tasks such as playing chess or GO. Thus, the statement we are making here is that significant contributions from other methods are needed beyond just combinations of depth and gradient-based learning. The statement is being made in objection to those who point at the human brain and say "look we [referring to our brains] are all neural-nets so we should just build bigger neural networks". Needless to say, this is a terrible argument. The arrangement of neurons, modules of neurons, all the other stuff inside the brain, and the processes and principles by which they all work together is the helpful thing to know, not that our brains are "all neural-nets".

In summary, the challenges are known and enormous, it is clear that depth alone cannot do all the heavy-lifting, rather it should be viewed as some basic necessity and too far from the sufficient requirements or features. For that reason, it's unlikely that we'd still call the future systems "deep learning" systems. It's also unclear how far away that future is exactly.

---

[139] ReLU (one of the most popular and biologically inspired nonlinear activation functions used in DL today) is of course non-analytic at the origin (when input is zero) and is being used quite successfully. However, in the practice of gradient-based learning this non-analyticity is simply ignored (even though it implicitly influences the learning) and has nothing to offer to help with discovering non-analyticity in target functions being learned. Yet, in mathematical theories that point of nonlinearity is the exact point you want to expose and form your theory around.

[140] Here's a deep network "doing math": G. Lample and F. Charton. "Deep learning for symbolic mathematics." arXiv preprint arXiv:1912.01412 (2019).

Hybrid Symbolic and Sub-Symbolic Systems

Obviously, we don't yet know what these "sufficient features" are, in order to overcome all the challenges we mentioned. In the meantime, given that many of the shortcomings of deep learning systems are among the capabilities of symbolic AI systems, many have been considering hybrids of symbolic and sub-symbolic systems.

We already talked about distributed representations. They are also referred to as subsymbolic representations because a single object or concept is not represented by a single symbol (e.g. a single number or word), but instead, by a collection of things as in a vector. So neural networks and deep networks, in general, are referred to as subsymbolic in contrast to the traditional symbolic AI. The symbolic-subsymbolic hybrids are also known as neuro-symbolic systems.

The distinction between symbolic and subsymbolic has old roots in cognitive science where it was customary to distinguish any model of cognition as either symbolic or "connectionist". Connectionism was in fact another name for neural networks (and still is within cognitive science and philosophy of mind), where the core belief is that cognition and mental states just arise from the connections of all the neurons, as opposed to being a property of any individual element like a neuron. Whereas in symbolism, a mental state has an individual representation and identity.

This categorization spilled over to the engineering of AI systems. As we already discussed, in the early days of AI, people were much more optimistic in replicating human intelligence, and a huge effort was spun on coming up with computational architectures that can perform many tasks holistically as the human cognition does. These architectures are therefore known as cognitive architecture and must have some degree of generality and broadness in their scope to qualify as cognitive architecture, or if not functionally broad, they must be compatible with some model of cognition coming from psychology or cognitive science.

In their early days, think the 70s and early 80s, all models for cognitive architectures were either based on symbolic models of cognition or connectionist models. There were no hybrids. As we saw with the advent of modern AI, handling uncertainties and probabilities became much more important. That gave rise to probabilistic logic frameworks and Statistical Relational Learning (SRL), the field which attempts to combine logic, probability, and learning. The name "relational" here just emphasizes the expressibility of logical relationships as in first-order logic.[141] Activities in these fields and modern AI (to handle uncertainty) happened to give rise to many hybrid cognitive architecture models.

It was as true two decades ago, as it is now, that it remains an open question how to properly integrate knowledge representations expressed in any symbolic form, such as with first-order logic, with statistical learning and inference techniques. The former often goes under the name knowledge representation and reasoning (KRR) and the latter obviously goes by the name machine learning. Such integration is exactly the challenge that SRL takes on, and among many approaches to SRL, inductive logic programming is a popular and general one.[142]

Now, why are such hybrid systems important and attractive? This question was addressed by Turing award winner Leslie Valiant already in 2003:

"The aim here is to identify a way of looking at and manipulating commonsense knowledge that is consistent with and can support what we consider to be the two most fundamental aspects of intelligent cognitive behavior: the ability to learn from experience, and the ability to reason from what has been learned. We are therefore seeking a semantics of knowledge that can computationally support the basic phenomena of intelligent behavior."[143]

Neural-symbolic approaches can be considered a subfield of SRL, although they differ from all traditional approaches within SRL, and have now grown to dominate the field of SRL and the world of hybrid systems, thanks to the successes of deep learning over the past decade. Research on Neuro-symbolic hybrids has been ongoing for more than two decades now.[144] And lately, they are getting more momentum.[145]

The hybrids may come with different names such as neuro-symbolic, neural-symbolic, symbolic-subsymbolic, relational neural networks, and so on. They all mean the same thing and they are all facing the same challenge of integration. The real differences among hybrids originate from

---

[141] Although it explicitly refers to "relational logic" which sometimes is philosophically categorized differently than predicate logic, in practice all relevant relations in these systems can be expressed via first-order logic.

[142] See Kersting, K.. Inductive Logic Programming Approach to Statistical Relational Learning, IOS Press, Incorporated, 2006.

[143] L.G. Valiant. Three problems in computer science. J. ACM, 50(1):96–99, 2003.

[144] See the 2002 book by Artur S. d'Avila Garcez, Krysia B. Broda, Dov M. Gabbay, titled "Neural-Symbolic Learning Systems: Foundations and Applications".

[145] Artur d'Avila, et al. "Neural-symbolic computing: An effective methodology for principled integration of machine learning and reasoning." arXiv preprint arXiv:1905.06088 (2019).

the types of integration between the symbolic and the sub-symbolic learner components, as well as how these components exchange knowledge. The integration architecture can range from not coupled or very loosely coupled to very tightly coupled or integrated to the point that barely any functionality of the hybrid system can be performed by any single component. As far as how they exchange knowledge, there are different ways to translate the work of the different components between them. Given that KRR and ML are at the core of neuro-symbolic hybrids, they always need some translation layer in between even if it is somehow implicit. In a very tightly coupled system, the work of the components would have to be translated back and forth many times in order to respond to a single function-call of the hybrid system.

When we talked about combining the ML methods, we mentioned the resulting ML system loses interpretability because of the combination. Well, in neuro-symbolic hybrids, the interpretability job is handled by the symbolic component as it is its strength after all. But obviously, in doing so, we have just turned one problem into another problem rather than solving it. The new problem is that we have come up with a good translation layer from the sub-symbolic component into the symbolic one. Early work in neuro-symbolic hybrids started slightly in the reverse direction. That is the case for systems known as Connectionist Inductive Learning and Logic Programming (or CILP) systems, where feedforward neural networks are actually there to implement a set of logical expressions. That is built on the foundation that for every logic program there exists a neural net that yields the same results, i.e. set of logical relations.[146]

There are many ways to combine the symbolic and subsymbolic units, which is a big world of its own, but the overall goal is to achieve both learning from examples and reasoning capabilities under one system. That essentially requires having some means to insert background knowledge into the learning process, extract symbolic knowledge from the subsymbolic units, fine-tuning knowledge of the system with feedback, and perform lots of deductions in parallel. All are features that neural-symbolic systems are set out to deliver in one way or another.

So how far can we go with hybrid systems to address DL problems we mentioned in the last section? As we said there, it is accepted that the flexibility and systematic frameworks that exist in most programming languages and classical AI in support of (system 2) reasoning and knowledge manipulation is missing from neural nets. That includes manipulating and exchanging small chunks of knowledge and systematic generalizations. However, the price we pay for the often ad-hoc design

---

[146] For more details see the 2009 book by Artur S. d'Avila Garcez, Luis C. Lamb, Dov M. Gabbay, titled "Neural-Symbolic Cognitive Reasoning".

and implementation choices in the hybrid systems may be too high. That is, although the hybridization can bring benefits, the symbolic units may become a source to reintroduce the same bottlenecks rule-based AI had to face, including non-scalable feature engineering and hard to encode human knowledge. Therefore, we need to make sure all the gains we have had in deep learning using depth and width (distributed representations) are used for reasoning as well. An example of doing so is somehow grounding all the symbols in some learned distributed representation so that the AI engineer wouldn't have to choose a name for every symbol. And if a name is given to the analogue of a symbol its semantic is declared by the underlying distributed representation.

Meanwhile, the answer that DL researchers have is the same old answer that we should represent everything by learnable parameters (from examples) and let the network make all the decisions. Now, the aspirations and the scope have grown a lot further. A great example here is the "differentiable neural computers" (DNC), where the idea is to add external memory units (holding distributed representations for long-term) to the architecture to show that every computer program can potentially be implemented also by a differentiable program, provided that enough examples of the given task are available to learn from.[147] That means symbolic reasoning and in general any algorithmic tasks could be emulated by DNCs or other fully differentiable programs equipped with long-term memory components so that various data structures could be represented, maintained, and manipulated.

An example of such algorithmic tasks is the good old array-sorting problem, which in most programming languages we can implement using a few lines of code (at least the not crazy fast algorithms for sorting). We can use a DNC not only to implement a sorting algorithm but also many other algorithmic tasks without writing a totally new algorithm.[148] However, for each new task we would need a different set of examples (millions or billions of them depending on the task) and to train the DNC from scratch, not to mention potential architectural changes or sensitivities to choices of hyperparameters. Therefore, although the prospect of neural computers is a very exciting one and a path to real innovation, it doesn't yet constitute any breakthroughs in deep learning. That is, they suffer from the same problems that deep learning currently does, many of which we reviewed

---

[147] A. Graves, et al. "Hybrid computing using a neural network with dynamic external memory." Nature 538, 471–476 (2016).

[148] This idea was first exhibited in Neural Turing Machines: A. Graves, et al. "Neural Turing Machines." arXiv preprint arXiv:1410.5401 (2014).

in the previous two sections. In sorting, for example, the DL systems can in no sense be said to exhibit any understanding of the task, rather what is happening is more along the lines of sophisticated curve fitting in very high dimensions.[149]

Another example of a fully differentiable system to emulate features of symbolic systems is in the domain of reinforcement learning. "Relational reinforcement learning" used to refer to reinforcement learning algorithms that take advantage of representations of entities, states, and actions in predicate logic to get many benefits of knowledge representations and reasoning of classical AI. Now, in the era of deep learning, for scalability and autonomy reasons (away from the explicit encoding of knowledge by human engineers), the deep learning community has been converting those ideas into fully differentiable versions. An example of such work is "relational deep reinforcement learning" in which proxies to symbolic relations are represented and learned by the network.[150] The idea is to design architectural inductive biases into the network in order to facilitate gaining relational (first-order-like) knowledge, such as entity x is to the right of entity y, by some distributed representations of the two entities and the predicate "being-to-the-right-of".

These examples, DNC, deep relational RL, and many others are obviously not hybrid systems but attempt to offer the same benefits all on their own. They are instances of a very hot domain in DL research to eliminate the need for hard and expensive-to-engineer hybrids that, even if successful, may not work in every context. Having said that, and as inspiring, ambitious, and exciting such activities in DL research for abstract and symbolic reasoning may be, they are still simply too far from what a programmer can do with a computer and powerful symbolic languages. Therefore, it shouldn't be surprising that the very best commercial AI systems out there today are hybrid systems that combine deep learning's strength for perception with some abstract rule-based or model-based reasoning.

In conclusion, as far as research on reasoning goes, there are two general directions. One is designing hybrids that attempt at general intelligence. They are cognitive architectures that remain on the more ambitious side of AI and cognitive science research, rather than commercial hybrids that can be functionally quite narrow. Another direction is, pursued by most deep learning researchers, to emulate all the useful features of symbolic AI in some subsymbolic fashion.

---

[149] Visualize it as point-by-point geometric morphing, using a lot of dense sampling from the input-output space, where "dense" means staying close to the right manifold representing the tasks.

[150] V. Zambaldi, et al. "Relational deep reinforcement learning." arXiv preprint arXiv:1806.01830 (2018).

Probabilistic Programming, a New Game in Town

A promising middle ground (though still in the direction of hybrid systems) is being recognized as "probabilistic programming". It is indeed a new game in the town of AI. What is it?

Recall two things. 1) In our discussion of "limits of DL", we mentioned a major category of objections to DL comes mainly from cognitive scientists who point out the differences between how humans learn and perform tasks, and current DL algorithms. 2) We also just said that fully differentiable systems are still far from the abstract reasoning that we can explicitly code for in most programming languages. So what about bringing the best of all worlds together?

As the name probabilistic programming suggests, we can combine the expressive power of regular programming languages to perform symbolic reasoning (including any formal logic), with probability models to handle uncertainty. It also allows us to combine any background or prior knowledge we have with any knowledge that can be learned from examples. The trick is to include all uncertain or learnable variables in an overall joint probability model that we can condition based on context, sample from, and so on. That would be a probabilistic program. Turns out this combination is a great computational framework to model cognitive processing as well: to learn lessons about cognitive psychology that could guide our AI algorithms.

So is that really new? Haven't we had a combination of logical reasoning and probability models before? We indeed have. As we discussed in the section of going from classical to modern AI, the onset of the transition to modern was in fact about bringing uncertainty handling capacity to previous AI systems which gave rise to technologies such as probabilistic knowledge-based systems. What's new now with probabilistic programming is that the expressiveness comes from regular programming languages rather than expressivity in any formal logic. So whatever we want we can in principle code for, as long as we can build them into an overall joint probability model represented by the main program or routine of the probabilistic program.

The other thing that is new is that we can have at least parts of the program learn from examples, whichever part of it that we want to be learnable. That means we can have differentiable programming inside probabilistic programming! Basically replacing some function in the program with a deep network that estimates it using examples.

Probabilistic programming languages (PPLs) provide native support for writing such probabilistic functions/operations. They provide the engine that helps us perform sampling, conditional probability calculations, implement Bayesian updating, etc. That is a PPL. What PPLs try to do is to take care of the inference part for us so that the programmer's responsibility is reduced to specifying the model and not implementing the inference algorithms on the model!

DARPA's program to fund probabilistic-programming research was launched in 2013. Since then support for PPLs has increased and there are many ongoing "PPL" projects active today. There is not yet any "most-popular" PPL that is heads and shoulders above all others. The reason is simple, fast and general probabilistic inference is an unsolved problem. Each PPL has choices to make as to which inference algorithms to implement and what kinds of speeds and accuracies should get priority over others. Regardless of where the main strength (what kind of inference and updating) points of a PPL are, they all do some heavy-lifting for you. As an example for the interested reader, I suggest looking into PYRO[151] that is a well-documented probabilistic engine.[152]

Another difference between probabilistic programming and the old logic-based systems with some uncertainty handling capability added-on is as follows. Recall, from our discussion in the reasoning and logic section, the two distinct uncertainties: existence-uncertainty and identity-uncertainty (uncertainty in what exists vs. uncertainty in the true identity of what exists). A prerequisite for handling these uncertainties in the traditional systems is to come up with all the symbols we need and to give them all distinct names. In the probabilistic programming world, that can be less tedious. That is, we don't have to give a name to every possible object, instead, we can just give names to the probability models (and their components and features, i.e. sub-models) of the object. We can specify some prior knowledge about the objects, environments, or relations, let the PPL bring in the data/examples to learn from and update our model, and then using a query given some context we can get to a particular instance, a unique object, to be called whatever we want then, not a priori. In this fashion, both existence and identity uncertainty is taken into account less explicitly.

There are other appealing aspects to probabilistic programs too. We mentioned the ability to bring in prior or background knowledge. That deserves more digging. We also mentioned we can learn from examples, but we choose what parts are to be learned from examples and represent other forms of knowledge, such as background or prior knowledge, separately in a way that the

---

[151] https://pyro.ai/

[152] To learn more about probabilistic programming, in general, I refer you to https://probmods.org/: Probabilistic Models of Cognition by Noah D. Goodman, Joshua B. Tenenbaum & the ProbMods Contributors at Probabilistic Models of Cognition - 2nd Edition.

combination could support probabilistic reasoning. That means, for instance, we can bring in our knowledge of physics of an environment, or other scientific models we have for how something in the world works, and let the model only learn about the specific context and statistics of the target problems instead of trying to figure out how the world works for which we may not have sufficient data or the right ML models to learn well. This is huge. The ability to insert custom knowledge in the middle of some giant ML system to assist with learning and removing some underlying uncertainty for the model results in the ability of probabilistic programs to effectively learn the same task from less data. Probabilistic programs can learn a probabilistic representation of abstract entities or concepts with any mixture of differentiable or non-differentiable components.

They may be able to learn not just from less data, but learn more from less data! By that I mean, they can learn rich representations that are more suited for abstraction and reasoning. Let's see how. Based on findings in cognitive science, we know humans store some kind of template for concepts or categories and then, based on a given context, they can generate/imagine new instances. That is essentially done by recombining more primitive templates in slightly new ways. Similarly, in probabilistic programs, we can reuse the template (the probability model) representing a concept into another program, just as we do in calling a method or library in "regular" programming. We can systematically adapt to many other tasks by reusing the abstraction that was previously learned as a probabilistic template, and flexibly modify it to get an instance for the new context, even though the new context would constitute a new distribution for the data and we don't have such data (that is an instance of out-of-distribution generalization). The way it works is that the new context provides the value of certain variables which we can use to condition the probability model representing the template and get a conditioned sample out. This sample would be an instance of the concept in the new context and that is exactly what we want.

Another appealing feature of probabilistic programs over end-to-end differentiable/deep learning systems is that they allow us to explicitly take control over what things need to be kept track of. For AI agents, it would be having some control over what the belief state of the agent should contain. This is an important feature sometimes necessary for many mission-critical tasks. Take the example of self-driving cars, obviously no matter how accurate and impressive their sensory and perception systems become, they should still keep track of things that cannot be seen or sensed. Rather, decision making should be based on high-quality belief states about what's true in the world, to which the sensors hold only a partial exposure. For instance, the presence of pedestrians or smaller vehicles behind bigger ones cannot be seen but could be inferred based on little clues in the subtle changes around, such as in the driving of the bigger vehicle or the behavior of other visible entities around. Furthermore, "keeping track" doesn't mean being aware of everything at all times but having a probability over it and keep updating it based on evidence, and that's what PPLs allow us to do natively.

For all these reasons, probabilistic programs have shown a great promise at contextual adaptation, the third wave of AI systems (yet to come) as DARPA puts it. PPLs are already in production use for various real-world applications for assisting scientific discovery to municipal predictive systems out there. However, penetration to the high-tech industry at large is currently very limited. One reason for that is, the openness and flexibility of probabilistic programming is also its curse. People have to actually write programs with the right probabilistic structure, and most programmers are not used to variables that represent probability distributions. Forming proper probability models requires mathematical and statistical expertise (and ideally cognitive science insights) beyond what's possessed by most engineers. There are tools and techniques that can alleviate this to some extent, but we don't yet know how to reach the gold standard that would give in for widespread adoption.

Chapter 8

AI-Labeled Economy and Data Science

So far we've been discussing the methods that are being labeled AI, their shortcomings, and what we currently have at our disposal to deal with those limits. Let's now turn our attention to where these methods are being used in the real world. That's all industries, not just the high-tech sector. Everywhere that there is some degree of digitization, there has also been some degree of AI applications. Let's discuss how the industry at large views AI and then some of the challenges with that "on the ground".

AI-Label in the Industry

If the research and academic community doesn't have it together in terms of naming practices, imagine how the industry would butcher the terminologies. No phrase with a marketing value whatsoever under the sun has been spared. Be it, applied AI, augmented AI, assisted AI, autonomous AI, cognitive computing, general AI, better AI, faster AI, on and on. If the research community could give away false promises, imagine what companies would do, whose survival critically depends on their marketing and messaging. We human beings get more excited about a product when it becomes more personal to us. So if the messaging around the label AI is inclined toward relating it to human intelligence, not only the message is understood easier (as we can relate to it easier), it also sells better. But there is no stopping the show now. Almost everyone is so quick to tell you some success stories on some AI project, product, feature, or a recent exciting research result, all resulting in mass refusal to think deeper.

Calling it AI and relating it to human intelligence is not the only clever messaging of the technologies that use data to predict stuff. Another way to make it personal is by having humans imagine themselves somehow paired with the technology. That is the case for the labels such as human-centered AI, human-in-the-loop AI, human-augmented AI, among others. They all try to message slightly different concepts but it all boils down to two things that they leverage. 1) obviously humans and machines have different sets of weaknesses and strengths, so we should be able to combine them for better outcomes. 2) for most applications the end target is humans, so we should make sure humans are actually benefiting by putting them at the "center" of the process or design.

Having said all that, the industry is not a single person, it has its own collective mind, operational philosophies, and beliefs with dynamics that are beyond the control of any individual. The reality is that for rather obvious reasons, industry loves *instrumentalism*. That is, a new technology like "AI" is practically perceived as yet another tool. Sure it may change how we have been doing some things in the past but it is in principle no different than other major technologies, such as databases or the internet (even the comparison to the internet gives the current AI-label somewhat of an elevated status). That's the main philosophy followed by the industry sometimes consciously and other times subconsciously.

What shouldn't be surprising are the successes and excitements in the industry. Those are all justified. The reason is simple: A) we explained how deep learning has become a highly practical function approximation method. And B) functions are everywhere, which is a super important fact to pay careful attention to! A function (being a map from some inputs to some outputs) can, very broadly speaking, take us from the information we do have to information we don't have. Let's call that a prediction to be consistent with how the majority of industry views prediction, or AI models, as. Now, if you already have sufficient amounts of good data (examples of a map[153]) you can have good predictions cheaply. Why cheaply? To discuss price, we have to discuss economics.

---

[153] Talking mostly about supervised learning here which constitutes the majority of successes in the AI applications in industry to date.

The Simple Economics of AI

Following the philosophy of instrumentalism in viewing AI as yet another technology, the economics of it don't need to require any fundamentally new model either. So one can use the simple view of classic economics of supply, demand, and prices. We can simply ask the question of what changes in the era of AI. As we said, the industry sees AI as a machine's ability to predict. Therefore, the answer is simply that the supply of "prediction" goes up! It is no longer supplied only by analysts or human experts. That means the prices of predictions fall and we can have cheaper predictions. Take the example of driving for instance. A human driver predicts what the right amount and direction of steering of the wheel, acceleration, etc. are to safely move or stop the car. What if the car, as the machine, could predict the same things? "Cheaper prediction" would mean that driving could become a lot cheaper. In many cases, where the prediction without AI is not possible, the price of prediction is effectively coming down from infinity to some affordable value.

When the price of something falls, the price of things that are associated with it (its complements) increase. Coffee gets cheaper, people use more of it, and the prices of cream and sugar go up. In the case of AI, the complements are obviously the data used to build the AI; the expertise to build the AI and maintain it; the infrastructure for AI to operate it; other tools associated with building, running, and maintaining AI; the human judgment on how to best put an AI to use; the produced data and experience of an AI operating in the real world; the ability to act on the prediction, etc. among less specific complements such as electrical energy used by AI. The price of all these gets pushed up because the price of prediction gets pushed down by AI. Data being the most critical complement of an AI stands to become the most expensive of all, and that's why we often hear "data is the new oil".

While the prices of complements rise, the prices of substitutes fall. The substitute for AI is predictions by human experts, where their prediction could be substituted by a machine prediction. For a more detailed discussion of these views, I refer the reader to the 2018 book called "prediction machines: The simple economics of AI" by Ajay Agarawal, and others.[154]

So far this model would be equally applicable to a commodity like coffee as well as a sophisticated IT technology such as AI. Many economists, however, consider AI to belong to its

---

[154] Ajay Agrawal, Joshua Gans, Avi Goldfarb. "Prediction Machines: The Simple Economics of Artificial Intelligence". 2018.

own category just because of how widely relevant and broadly impactful it is to bring down the price of predictions. As we saw, in the case of self-driving cars, many problems that are not traditionally thought of as prediction problems can be turned into prediction problems. This is also not particular to the case of AI, rather, already captured by simple price economics.

When something gets cheaper we don't just use more of it in the ways we used to, we start using it to solve other problems or satisfy other needs that already have perfectly good solutions in place. Even if doing so may be more expensive in the short-term, long-term convenience can be a strong mind-shifting force. Take the case of batteries. Battery technology has been making significant progress recently i.e. becoming cheaper, and therefore we are now thinking of battery-powered cars, houses, and so on. Similarly, innovative AI applications are turning more and more problems into prediction problems. Think of front-end development and coding a single webpage. Once you show many examples of code and how the associated page looks like, the machine can predict what the code (or the code template) should be to produce a certain design or mockup that a web designer has sketched. That is not a great substitute, just yet, but one day it could be.

Given all these, what's clear is that it's hard to see the limits for AI. The mind can be said to be a prediction machine, a point well-articulated by Jeff Hawkins in the best-seller "On Intelligence". That excites economists understandably and challenges them to figure out what is to be anticipated in AI-dominated economies of the future. However, doing so based on the simple economics of AI and the prediction machine view of AI can fall significantly short of what the future may actually bring.

Problems With the Simple Economics of AI

The simple economics of AI we have presented basically views AI as a collection of independent prediction machines. There are only two factors or input variables into the study of such models of an AI economy:

1. How many problems have been converted into prediction problems and are replaced, or will be replaced, by a "prediction machine"?
2. What's the accuracy of those predictions?

That's it. That's the model, based on which, an economist can crank up each of those variables and see what are the interesting things that can happen. Regarding the first variable, we already mentioned that we should anticipate application of AI to many areas that may currently seem far-fetched. Regarding the second variable, the industry is super excited about all the new doors that sufficiently high accuracy can open. It may qualitatively change how we do things. For instance, if you can predict a customer is going to need something with sufficiently high accuracy, it makes it profitable to ship it to them before they order it and risk logistical expenses due to some relatively tiny fraction of error. A level that a company like Amazon would aspire to get to.

As exciting as these projections maybe, reality will perhaps not unfold in this way. The reason is that such simple economics is too simple of a model to be stretched this far. Linear thinking based on such models will only divert us from recognizing the issues we are to bump up against first. Let's see why.

We can acknowledge that the simple model of "prediction machines" is fine for the time being, when supervised deep learning is dominating the AI industry (in the short-term), but if we are to make long-term projections and freely turn the accuracy knob super high in our imagination, there are a few assumptions behind the simple model which we must break. We'll cover two of them, as follows.

1. False Independence Assumption From Decision Making

The simple model totally separates prediction from decision making. The view is that we can simply use better and cheaper predictions as input into decision-theoretic (or utility-theoretic) frameworks and proceed with business as usual. The fact that decision theory has existed since the mid-last century, certainly before modern AI methods were developed, is perhaps the origin of such misguidance.

As we discussed in chapter 6, decision making and predictions are much more deeply intertwined than our basic intuition may make it seem. The independence assumption will only result in toy models. It may not be possible to independently raise the accuracy of a collection of predictions in a dynamic world beyond a certain threshold without deeply involving AI with decisions, actions, and their quality beyond just the accuracy of a single prediction. So let's drop the assumption that a decision is something separately built on top of predictions. Initially, it will be so, but won't be so for the fancy imagination projecting far ahead. What do we mean by initially and how long does it last? What's the time scale here? The answer is about a decade (meaning it's not a year nor a century) and that should get clear by the end of this section.

2. Future Scale Issues With Current Deep Learning

The simple economic model is mainly guided by the successful examples of supervised deep learning (DL) projects in industry. As we said earlier, we shouldn't be surprised that DL has been successful in taking the world by a tsunami because FUNCTIONS are everywhere and DL is suited to approximate them well. What I want to tell you now is somewhat the opposite, the other side of the same coin. The current practice stands to fail, in the long run, for the same reason (with a shift in emphasis) that functions are EVERYWHERE!

If it is not immediately clear what I mean by "functions are everywhere", please read appendix B on sets, functions, and maps. In that sense, there are an infinite number of functions "on earth" that we may care about, the knowledge of any of them could help with operations of some entity, business, or government. Are we going to try to collect datasets for each and approximate them all if a case for the ROI could be made? How long and how far would we be able to advance on this route?

The industry is already a few years deep on this path and still, there are plenty of low-hanging fruits remaining to keep us busy for at least a decade. Not to mention that there are many kinds of functions we don't yet know how to approximate well which are currently under heavy deep learning research, as we discussed in the section on "extending the methods". These and many more new developments in AI research, are expected to come out and shift the course in industry, albeit slowly. For the sake of this discussion, we are dismissing future development and considering only the current practices. Let's also completely put aside work in robotics and embodied AI agents which these arguments don't quite apply to, nor does the simple economics model. For context, let's think more along the lines of enterprise and industrial use cases addressed by supervised prediction models.

With that said, the question is: does this approach of going with an ever-increasing collection of independent prediction machines really scale? or maybe these functions shouldn't be considered independent, maybe their interactions are key to making optimal decisions, and what about the function that we fail to consider, approximate, or gather data for?

Let's take a look at the past. Recall expert systems? The problem with expert systems was that they wanted to hand-code a lot of features and facts into the system. To get it to the level of performance the industry has set its expectation on, it proved to be unmanageably too much labor by knowledge workers. So it didn't scale. Now, could it be that we are going to have the same problem with function specifications (including dataset and objective function engineering) in the current era of deep learning? In the sense that we need to do that for every case, every interesting pair of input-output data, and yet there are too many relevant things we are ultimately interested in to the point that these functions start to overlap and interact consequentially.

Of course the situation isn't quite comparable to expert systems and people argue that in practice the things we are interested in are quite small: you want to diagnose eye disease, give me whatever data you have on that as well as what you want to predict, and we are good. Perhaps in such cases, which we can call "static cases", the data and what you are interested to predict, all remain valid for a relatively very long time. But try running everything in a highly complex environment like a dark factory (a factory with no humans in, and therefore no lights on) with currently celebrated AI methods. You are bound to find yourself in a giant spaghetti bowl made of all the individual models and functions you have approximated. Potential unmanageability of this complexity aside, before long, the bottlenecks in running this environment have conceptually shifted simply because this factory is a dynamic and open system, whereas your functions were built upon many static assumptions about your data, metadata, application logic, so on and so forth. Whatever strategy you come up with may expire soon, and sooner than you can manage to react and modify the spaghetti bowl that is supposed to run it all optimally.

This would be a new regime that AI applications would have to enter and we are emphasizing that collections of prediction machines don't seem to be up to the task. That is, we are bound to hit scale issues with these practices. For accuracy to go really high, to the level that the economists such as the authors of "predictions machines" like to crank it up to, we most likely need to go through this new regime first, before the imaginary novel economic phenomena have any chance to emerge. Finding ourselves in that new regime, we'd have no choice but to integrate many models and prediction machines, and allow many decisions to be made in deep concert with one another, balancing many tradeoffs. For that reason, the simple economic model of prediction machines cannot be extended too far into the future.

Study of the interactions among different models, functions, and applications in this new regime was in fact the subject of an AI startup I co-founded in 2018. We knew the large impact considering the interactions among prediction machines could have on any business as a whole. But in the world of startups timing is everything and we weren't and still aren't in the era of mutually interacting AIs yet. For the time being, these matters (putting models together) are simply left up to the so-called application logic layer of AI software products.

We mentioned that the time scale to step into the new regime in industry, is roughly a decade. That is simply due to the nature of the opportunities lying ahead, as well as the challenges the industry is already facing. As for the opportunities, the current methods can take us really far on their own, which is why the mindset in industry is aligned with that of the simple economics model of prediction machines. Regarding challenges, let's mention two things that can't be addressed over the course of a year or two, and thus are decade-long pursuits. These challenges are much more immediately occupying than the future scale issue with functional interactions we have been discussing.

The challenge of digitization and big data:

Recording digital footprints of all operations is still not a given in 2020s, with many pockets of industry (manufacturing, logistics, and many parts of health care) lagging behind. Of course industry leaders in various sectors don't represent the industry at large. Even in industries that are almost fully digital, like the internet industry, we find other challenges regarding their data. For the past decade or so, we have been talking a lot about the challenges associated with big data, which as we mentioned in the first chapter, the big is not necessarily about the volume of the data, but also its velocity, variety, veracity among many other factors. Inventing and developing suitable technologies which store and enable working with such big data, the so-called big-data technologies, and implementing them till adoption across various organizations have been and still are far from trivial. We have seen migrations from traditional data warehouses to so-called "data-lakes", and recently "lake-houses", along with many other technologies that are typically served on the cloud. Moving data *consistently* around when you have a lot of data, with many models on top, and many people to serve, are matters that will keep occupying us for at least the coming decade.

Putting these basic issues which are massive engineering challenges on their own aside, there are other features we desire from big-data technologies towards the end goal of better serving analytics or AI applications. An example here is technologies that focus on surfacing relationships in the data, such as graph databases. These databases make the graph structures, i.e. the entities and relationships among them, first-class citizens in the way that the data is stored and queried. Given that most existing knowledge in the world naturally comes in the form of a graph, or can be represented as such, we've been seeing a little resurgence of graph-native technologies in the past few years and are likely to see more of them in the future in more advanced forms.

Challenges around the obvious scaling issues in terms of sufficiency of data, compute, talent, and research study transfer: The need for lots of good data and computing resources is no joke. Both are highly expensive. Let's not forget that in current deep learning, life starts at a billion examples and the models are getting increasingly larger, requiring much more computing resources for every step of the way from R&D to serving clients. Of course, you don't always need to have your own data but someone must, in which case, you might be able to use their model as a starting point and be able to tune it to your problem with a lot less data.

There is also another challenge that the industry has been silently dealing with. That is the challenge of transferring academic research insights and results to industry. Given the empirical nature of the majority of academic AI research, there always remains too much leeway in the practice of applying the promising methods to any real-world problem in industry. Going from an academic research paper to a successful AI product is extremely far from trivial and to do that, a best practice would be to start with some rigorous research focusing on transferring the academic research into the target circumstance. And this kind of position is not yet formalized in the industry, one that focuses on "research on practice-proofing AI research". These challenges are also decade-long issues to be addressed.

Data Science, a Missing Manual

The discussion of the previous section really sets the stage for us to discuss how to make sense of what "data science", still kind of a mysterious phrase, is supposed to be about.

By now, you should at least know and expect for ML to be a big part of "data science", regardless of how we may define it. Per our earlier discussions, ML is basically about the ability to improve some performance simply by providing some examples, some data. So we may find it natural to say, " well, to figure out with what data, how, when, and why, we need some "data SCIENCE", what else are you going to name a practice as broad as this." As tempting as it may be to be content to stop it here, we need to get a lot deeper than that. The reason is that even our definition of machine learning itself is overly broad. If we want to get squared with what data science should mean, we need to first get a little bit more clear about the practice of machine learning!

In fact, we should go even a little further back to statistics. Isn't statistics the field that most people think of concerning a discipline to deal with data? Isn't statistics some kind of a "data science or science of data" to most people already? Here's a fun quasi-fact; if you want to see a fistfight break out in one of your meetings with people who are mixed statisticians and machine learners, make a statement about machine learning being part of statistics or vice versa, God forbid. So definitely our topic here can be a sensitive one but we got to talk about it anyway.

Let's take a step back and ask why do we even work with data, what are our end goals? You may say, well, to prove or disprove some hypothesis, gain some insights, be able to predict something, on and on. Yes, but ultimately it's because we are agents in this world and we have to act in it, which means we have to constantly make decisions that precede any action. So it all boils down to decision making, a concept that as trivial as it may sound, deserves a lot more philosophical as well as scientific work. Hey, didn't we say we have some decision theory in place already? Yes, but that well-understood decision theory is just about one decision of one agent given some predictions and utility values for the options. That's nothing compared to what the science of decision making and acting need to get to. The world is full of many agents (living or not, natural or artificial) simultaneously making a set or sets of decisions. Moreover, as we said in the last section, you cannot always think of predictions and decisions as separable, especially when complexity rises. That unquestioned assumption of separability is made only by the "prediction machines" view of ML and AI. An insufficient view, as we argued.

Statistics in the modern sense started in attempts to use the data gathered from citizens for governance purposes, things like census planning efforts. Starting that way in the 18th century and then slowly percolating to many other fields by the 19th century. Of course, formalization as an independent discipline happened only in the 20th century just like many other fields such as computer science or electrical engineering. Now, in order to have an opinion about where it would go next in the 21 century and later, one has to have some beliefs about the ultimate goals and the current maturity status of the field relative to the goals.

Let's remind ourselves that the position we're taking in this book is set by our philosophy regarding the evolution of any discipline. That is to metaphorically imagine there is a missing manual that nature holds and it contains the ultimate philosophies, truth, useful facts, and recipes regarding the subject under consideration. And that our job can be interpreted as putting together an artificial version of the "natural" manual that is missing, and to gradually get the artificial manual closer to the "natural" one. Think of the natural one as what God would put together and the artificial one being our collective knowledge at any given time. We also swapped the word manual for the more suitable MANUON (manual of nature or natural manual) in order to drop some of the inappropriateness that the word manual carries in its literal sense.[155]

Alright, let's ask that if our knowledge of statistics is some kind of an artificial manuon, what would the corresponding missing manuon contain? Don't we already have all these powerful theorems and universally abundant statistical distributions of natural phenomena, etc. Aren't some of those Math theorems, universal power laws, eventual Gaussianities, fat-tail Poissons, already our visible paintings on top of some of the invisible strokes of God's pen?

Well, perhaps some of them could be but we have barely scratched the surface. Why? We mentioned already the complexity that increasing the number of decision makers can introduce, and that we don't yet understand the nature of such complexity, let alone figuring out what we can do with or how to engineer around it. There are many other complexities like that. Try increasing the number of decision sets for each decision maker, try increasing the time scale involved in decision making processes unboundedly, try increasing the length scale for decision making processes (think planetary scale and one day even beyond), try raising the stakes on going with the wrong choices of actions or inactions, on and on. We don't have much, if anything rigorous at all, to say about any of these phenomena. We are simply not there yet. Nevertheless, our technological advancements have already made it possible and profitable to build systems that can experience natural phenomena in these regimes that we don't understand quantitatively, without any accompanying ability to properly measure associated risks in the short or long run. Social media has been one hot example. Search and recommendation engines are another, though not unrelated.

In his 2018 blog post "the revolution hasn't happened yet", Michael Jordan, a world-leading statistician at UC Berkeley, who has trained many well-known ML or AI researchers, writes:

> "… just as humans built buildings and bridges before there was civil engineering, humans are proceeding with the building of societal-scale, inference-and-decision-making systems that involve machines, humans and the environment. Just as early buildings and bridges sometimes fell to the ground—in unforeseen ways and with tragic consequences—many of our early societal-scale inference-and-decision-making systems are already exposing serious conceptual flaws." [156]

---

[155] Quotes the summary in Appendix A.

[156] https://medium.com/@mijordan3/artificial-intelligence-the-revolution-hasnt-happened-yet-5e1d5812e1e7

So with regards to the science of data and decisions, what we currently have is some kind of an insufficient artificial manuon. Several fields are contributing to the existing draft. If you are working in a statistics department, you'd simply call your contributions: statistics, statistical principles, theorems, or methods. If you are working in a computer science department, you'd call your contributions "machine learning" or "AI". It's not just a matter of difference in the names or even techniques, the goals and problem definitions are somewhat different too.

Statistics, which is much older than machine learning, is clear about the goal being to measure and reduce the risk in our decision making and has more of an explicit focus there. In that sense, goals of statistics are more aligned with the scope of the full manuon we ought to be seeking. Whereas, the ultimate goals of machine learning are perhaps more confusing. Set aside the broad and formal definition of machine learning as it doesn't say much about the goals. Many computer scientists may tell you that the goals of ML are pretty much the same as the goals of AI. But we know that the mainstream usage of the label "AI" has been about "prediction machines". Therefore, we can conclude that ML has been mostly focused on writing or drafting that "section of the manuon" which is just about increasing the accuracy of isolated predictions. Statisticians point out the insufficiency they observe, and some computer scientists respond by saying "yea but you are not focusing as much on the cool prediction methods that we are after... let us figure that out and you can later put error bars on the outputs of our prediction machines!", in strong dismissal of the highly interdependent nature of decisions and predictions.

Putting these territorial and communal dynamics aside and who works where and with what title, there is a bigger picture to focus on here. And that the boundaries of all these fields, statistics, ML, AI are quite artificial as they are all part of a much bigger field whose goal can be compactly represented as finding a missing manuon with a scope big enough to encompass that of all those fields. Just to give it a name, let's call it the missing manuon of data and decisions. It is really as broad as that. That is what the industry is choicelessly facing. What data science has to cover starts before any data is acquired or produced, and then doesn't even end when the consequences of some of the decisions have returned. Data science is about the whole shebang, from better data all the way to better decisions in real-time, all the time, and in the open system of the real-world that is getting increasingly more complex. Except, that science doesn't quite exist yet! In other words, data and decision science, or data science for short, is still a missing manuon to be found!

Chapter 9

Manuons of Intelligence

Now that we have gone through this long journey of what AI as a label is and studying what people actually do under that label, let's raise some questions about AI beyond what people are currently doing, similar to the questions we just raised with decisions and data science. Specifically, we ask, if what people currently do under the label AI could collectively be viewed as an artificial manuon, what would be the corresponding missing manuon? Again, a missing manuon is by definition supposed to be the most imaginatively complete, consistent, and future-proof version of a subject, whereas an artificial manuon is some current proxy we have for it.

Even without knowing how to uniquely define it, "intelligence" is what we are ultimately interested in as a subject and as something we'd like to create and enhance. Thereby, let's assume the existence of a missing manuon of intelligence. Now, the question is, to what extent the things we currently have or call intelligence, or AI, can be considered the artificial versions of that missing manuon of intelligence?

It should be rather obvious that several subjects would all overlap with that missing manuon of intelligence. We need to carefully distinguish between at least the three biggies here, which are obviously AI (whatever the label stands for), human intelligence (HI), and decisions and data science (DD), which we just covered. This is no simple or clear-cut task due to the fact that they are all related and heavily overlapping, yet without any of them having been clearly defined much less well-understood.

Human intelligence is often referred to as "natural" intelligence. But we are going to be a lot more careful and not assume equity between human intelligence and what natural intelligence could ultimately be best thought of. In fact we already gave a name to such a more fundamental form of intelligence in nature. In chapter one, we discussed the idea of looking for a unifying framework for various approaches towards intelligence and reserved the name "Fundamental Intelligence" or FI for that unifying roof. Here we identify THAT with the missing manuon of intelligence we ought to be seeking. Of course the existence of this category of intelligence is a hypothesis on its own that we'll discuss soon. Mindful of this in the background, let's first try to separate out these overlapping manuons of intelligence we have mentioned so far.

Intelligence Versus Decisions and Data

Let's start with separating the manuons of DD (decisions and data science) from all forms of intelligence manuons. Per our discussion in the last chapter, DD requires and consumes intelligence. In fact, any manuon of intelligence could be considered a sub-manuon (or subset) of some manuon of DD.

For human intelligence, this is simple to see. In common wisdom, we often say intelligence alone isn't enough to succeed. Even for the most intelligent, it's not trivial to figure out how to acquire the right data and make the right decisions in the open system of the world and life. For AI, ASSUME we are already technologically there to fully imitate or even replicate human intelligence (whatever that could mean) with the difference that we are giving that intelligence a lot more resources, putting it on steroids, if you will. However, such AI is still one entity, by definition, whereas the subject of data and decisions we discussed previously spans way beyond just one intelligent entity and hence beyond the subject matter of an advanced AI. For a manuon of AI to subsume that of DD, it must cover interactions of such entities, the complex system arising from it, as a rather natural phenomenon, and how to engineer around it.

With regards to some ultimate science of intelligence or fundamental intelligence as we called it, we can draw the same verdict. As long as we constrain "intelligence" to be a property possessed by one agent or one entity (even if it is one emergent whole), it's only reasonable to keep decision making and multi-agent scenarios outside the domain of a single intelligence or intelligent entity. However, we must note that it is possible that ultimate notions of intelligence can render many data and decision making topics trivial. Nevertheless, they remain distinct topics. That is, intelligence remains as what makes data gathering and correct decision making easier, rather than being equivalent to them.

AI as the Current Artificial Manuon of Intelligence

We have discussed a lot of methods that people use and call AI, let's put them all in a box with the AI-sticker on. This box is our current artificial manuon of intelligence. Can we say anything about the anatomy of this manuon?

We saw it's hard to combine different sections of it as different methods are suitable for different problems. We saw deep neural networks are overriding many other sections of the manuon. We saw all sections may not need to get consolidated as the idea of a "master algorithm" desires. So the manuon is mostly a set of methods concatenated all together in no particular order, and we keep appending to it. A lot of what we discussed so far as AI methods can be considered to be part of the content table of such a manuon!

In that content table, we can see some clear separation between non-ML-based AI and ML-based AI including deep learning based-AI. Being the main force behind the buzz of AI as a label, deep learning dominates what keeps getting appended to the artificial manuon. We also saw that there are deep limitations with every one of the approaches as compared to human intelligence or general capabilities we'd like to have.

We discussed *generalization* as the central topic in machine learning, the core of statistical learning theory and modern AI, and PAC-learning as a basis for machine learning. We saw the strong assumptions or definitions that we start with there. Namely the following:

1. Assumption of stationarity (limitation to stationary distributions within which to seek generalization).
2. Having to know the hypothesis space for learning ahead of learning time.
3. Polynomial convergence especially in the number of examples to learn from as the bar to pass for all forms of machine learning.

With all these conditions in place, we can attain a generalization that is about one problem and one problem alone, only one application in one domain. This notion of generalization may ironically be even called "specific generalization". In that sense, we saw no "general generalization" in AI. We did mention "out of distribution generalization" but that is still far from a rigorous notion, in fact it's barely more than just a name for going beyond the current notion of generalization in ML or "specific generalization". On the contrary, we consider some form of "general generalization" to be a core feature of human intelligence. Though we should also note that we only anecdotally understand the nature of such "general generalization" in human intelligence and not yet with any mathematical theory of "generalizations in humans".

Alright, if AI with this ML basis is not to be qualified as really intelligent, or supposedly so different from human intelligence, how come we are told so much of the deep-learning-dominated AI is inspired by human intelligence? That so much of artificial neural nets are based on the neuroscience of brain neurons? More importantly, if we are not on-track towards surpassing human intelligence, how come we can already build machines capable of doing so much of what humans can? How come we keep getting closer to human performance (or surpassing it in some limited cases) in many domains that are believed to require intelligence, be it image recognition, driving, or playing ancient games like GO or complex video games?

Isn't human intelligence always represented by some ability to achieve complex goals? So how much does it make sense to separate human intelligence and AI?

AI & Human Intelligence: Are They Related?

We don't know yet how to best measure human intelligence or even how to define it well, we just know it's there, sitting pretty impressive and general. We don't know how to define some proper notion of generalization, commensurate with what human intelligence must represent. So it's currently hard to give convincing arguments that there is no intelligence in AI methods, ML-based or not.

On the other hand, machine learning with all those limiting assumptions is proven to be incredibly powerful and can achieve so much. Let's take a look again at those three machine learning assumptions we listed above. The first two, namely stationarity and prior knowledge on hypothesis space, may both very well be what the human brain also implicitly uses in one form or another. That is, we may have similar strong priors built-in by evolutionary design. This also means that whatever notion of generalization for human intelligence we eventually formalize, it cannot be a literally general generalization. A fact that is also absorbed in what we termed the fundamental bias of our brain. But as for the third criterion in ML, a polynomial rate may be fast enough for definitions of machine learnability but it may very well turn out to be too slow for ultimate definitions of intelligence.

Having said that, even with its cheesy name and naming culture, machine learning doesn't claim much on being no different than human intelligence. That kind of rhetoric is often associated with statements and projections coming out of the deep learning world. And that makes sense considering the origins of deep learning in the *connectionist* philosophies of mind. So that's what we want to discuss here. Digging a little bit into the nature of the relationship between human intelligence and deep learning.

We are not going to discuss the limitations of DL or that humans do things differently or more efficiently, none of that, some of which we have already covered and the rest has been pointed out by many cognitive scientists elsewhere[157]. Quite the contrary, we want to ignore the limitations and stay within the strengths and achievements of DL and even assume continued progress for it in the future. With that and the natural desire of most people to put human intelligence and celebrated DL-based systems into the same bucket of intelligence, we want to ask if it is fair to say that deep neural networks have anything to do with the brain? Is there any sense in which we can say DL is directly related to human intelligence?

The relation of DL-based AI to human intelligence is claimed on two separate fronts. One is based on how much artificial neurons are inspired by neuroscience and modeled after the biological version. The other is based on state-of-the-art performance of DL models (which in some tasks achieve or surpass human-level performance) with general and widely applicable methods like backpropagation. Belief in a direct relationship between DL and human intelligence moves many to seek biologically-plausible variants of the successful methods in DL-based AI or investigate how the brain may be implementing or approximating them. In the following, we are going to play the devil's advocate and argue against any evidence of a direct relationship between human intelligence and deep learning. We conclude that given our strong confirmation biases, it's best to assume no direct relationship between human intelligence and deep learning. Best in terms of making progress in finding the ultimate manuon of intelligence.

Obviously our lack of sufficient understanding of how the brain arranges and works with its neural network leaves room for too many possibilities. This has given rise to many old debates among philosophers of mind and cognitive scientists. Arguing against or for connectionism and

---

[157] Lake et. al. (2017). "Building machines that learn and think like people", Behavioral and Brain Sciences, 40, E253. Also see the 2019 book "Rebooting AI" by Gary Marcus and Ernest Davis.

discussing how to realize or implement various models of cognition.[158] That's where we have to start, though without getting into any philosophical weeds.

Although the phrase connectionism has its roots in 19th-century psychology, as an ideology it only started to gain substance in the 1940s shortly after finding an abstract model for a single neuron, which is often referred to as a McCulloch-Pitts neuron. The idea of connectionism is that once we connect up lots of neurons that are all processing different pieces of information in parallel, higher-level cognition and intelligence emerges. Computer scientists then borrowed the concept and, for a long while, parallel distributed processing (PDP) was a phrase in fashion for neural networks as special cases of multiple-instruction-multiple-data (MIMD) machines, where not only data is distributed but the parallel process on splits of data could also be different.

But unlike the architectural ideas behind any instance of MIMD machines, the idea of connectionism does NOT say anything about how the neurons are to be connected and whether each neuron has an independent functional identity or not. Regarding functional identities, there are ONLY two distinct possibilities then.

1. First possibility is inspired by the McCulloch-Pitts neuron model that we just mentioned, while making as general as possible. In this case, an artificial neuron or a fixed group of such neurons (with a fixed connectivity pattern) has a specific function.[159] Let's call this the McCulloch-Pitts framework. This is the framework that we studied in the deep learning chapter since it is exactly the model used in current deep learning neural networks, in which we start with a fixed architecture and each neuron can offer a specific function, for instance, it "lights up" when a specific feature is present or detected (e.g. we can have a single neuron that lights up only when a cat face is present, i.e. a cat neuron). Note the specific function does not need to be interpretable by us.

2. The other possibility is that only *assemblies* of neurons have functional identities, and each neuron can be a member of multiple assemblies.

---

[158] Buckner, Cameron, and James Garson, "Connectionism", The Stanford Encyclopedia of Philosophy (Fall 2019 Edition), Edward N. Zalta (ed.),
https://plato.stanford.edu/archives/fall2019/entries/connectionism/

[159] The original model by McCulloch & Pitts was strictly binary and non-differentiable as their mindset was around how can the brain support logical operation as we discussed in chapter 3. Here, we are ignoring that restriction as we are interested in the general framework put forth by them, not their specific work, i.e. the circuit design for a logical neuron which happened to be among the inspirations John von Neumann gathered to come up with the Von Neumann architecture.

Yet, no one knows if the intelligence in the brain is using only the former, the latter, or some mixture of both.

Putting the unknowns within connectionism aside, it all boils down to our understanding that "distributed representations" being one of the two pillars of deep learning (as we discussed in the deep learning section) are essential for information processing contrary to what the classical and purely symbolist models of cognition advocated.

Names and labels just reflect the dominant philosophies of any given time or era. We just need to be careful not to take them to be more than that. The bare idea of connectionism and being sub-symbolic, does not constitute a direct relationship between human intelligence and deep learning. A network of many processing units is not constraining enough as a framework, not constraining types of processing units, the arrangement of them, communication protocols, etc. In fact the idea of a neural network is so broad that it is hard to find models it could not represent. For instance, local models (i.e. models with non-distributed representations) such as radial-basis-function networks, self-organizing maps, or even non-parametric techniques such as Parzen windows[160] can all be implemented as neural networks.

So the word "neuron" barely carries any meaning, nothing more than what the phrase "processing unit" does. Meanwhile, many deep learning researchers when asked about their faith in deep learning's future often respond by saying "we [our brains] are neural nets all the way so we just have to make our artificial neural nets better...", in dismissal of any relevant difference between brain's neural networks and artificial neural nets.

We know intelligence is a higher-level phenomenon than what's captured by a single neuron. What we don't know is how many levels of emergent phenomenon there are before we get to intelligence or other cognitive phenomena. In volume two of this book, after we have discussed the nature of nature and mathematics, we are better equipped to discuss nested levels of emergent phenomena. For now, we need to recognize that going up by one level of emergence, can radically change the mathematical representation of the system.

Suppose we call forming a distributed representation by connecting neurons, the first emergent level. It's possible then that a second-level phenomenon could be realized by a separation, rather than connection, of distributed representations (we may interpret this as depth). Just like the second level is made of distributed representation, rather than single neurons, other higher-level

---

[160] Donald Specht, "A general regression neural network." IEEE transactions on neural networks 2.6 (1991): 568-576.

phenomena are not describable in the space of neurons or in the language of neurons. They are best described in terms of other mathematical objects that have emerged from a bunch of neurons, layers, or assemblies (of neurons), forming the building blocks for that higher-level description. The mathematical properties of higher-level building blocks and how they emerge are precisely what needs to be studied and discovered. Again, the point is that we may have a long way ahead in discovering all that before getting to see what intelligence could be best represented as.

We often hear "deep learning AI is inspired by the human brain", and there may be some truth to that but we must realize that it does not carry any weight in establishing a direct relationship between human intelligence and deep neural networks. *Inspiration doesn't warrant relation!*

Wait, what about the "similarities" between modern deep networks and the human brain? Given that we only have the language for neurons and connections of neurons, we are only equipped to discuss the lowest level phenomena. First off, to talk about similarities we are forced to assume, what in the above we called, the McCulloch-Pitts framework captures how the brain works at some basic level. Then, one could say modern deep networks have three basic features shared with the human brain. One is the model of an individual neuron, artificial neurons used in modern deep networks have some similarity with those used in computational neuroscience, specifically the elements of a non-linear activation function (often a ReLU function) on top of the sum of input stimulus. It must, however, be noted that real neurons feature a crazy amount of physiology, chemical and physical dynamics involved in their life, that are all absent from those used in current deep networks, even if we assume (under this framework) single neurons are performing a specific function.[161] Thus, this similarity maybe the weakest of the three.

The second similarity is what we called width or a distributed representation, which as we just mentioned is the main meaning captured by connectionism. The third is depth and we know that the brain also employs hierarchical structures for its information processing. For instance, depth in the early stages of the visual cortex is very well observed, exhibiting many similarities with modern deep networks in computer vision.

Does having these three basic features in common warrant a relationship between human intelligence and deep network? What if these features happen to be the same in both evolution-resulted brains and modern deep nets by coincidence? What if these features are set by general facts

---

[161] Koch, C. "Biophysics of Computation: Information Processing in Single Neurons" Oxford University Press, 1988.

in nature, the mathematical nature of information processing, or the nature of information sets we humans care about processing?

Consider the "width" feature. We mentioned the only way to tame an exponential complexity is by taking its logarithm and that's what distributed representations do. That's a fact of nature that any information processing system dealing with exponentially diverse data has to abide by, be it a human brain, a deep learning network, or any other system.

Consider the "depth" feature. We believe the world is made of things and things are made of other things, and it keeps on going that way. So far, that's the most compact mathematical description we recognize for the world. Thus, the raw information coming to any general information processing system, be it a human brain, a deep network, or anything else, is best broken down into multiple levels of abstraction to be consistent with how this information is generated or most compactly understood. Again, this is a fact of our nature that perhaps the most complete explanation comes from parts of parts of parts and on. The spirit of deep learning is to have repeated representations of representations in order to take advantage of this fact.

Consider a single neuron. Let's set aside all those physiological, chemical and physical complexities, and for a second assume that it's effectively computing just as a ReLU cell would. Now think of the simplest yet useful processing unit you can. What can you find simpler than summing the inputs, and outputting it if it's positive (as in a ReLU)? I can't think of any. This simplicity is yet another fact of nature, the mathematical nature.

Putting the above arguments together, the common existences of these features do not establish a direct relationship between human intelligence and deep networks. The fact that our historical path to arrive at our modern deep neural networks has been by looking into biology and being inspired by human intelligence, doesn't mean one couldn't have arrived at the same principles and models, without looking at or studying brains, or any knowledge of biological neurons whatsoever. A few bright fellows with the right philosophies, abundant computational resources, and sufficient data, could have!

Notional similarities to the human brain and intelligence are not limited to neural networks. In fact every other AI method we have covered can be claimed to bear some similarity to the human brain or intelligence too, in one form or another. Though they are all similarities at a higher level of cognition as opposed to the biological or neuroscientific level. Think of the methods in the analogizer camp of methods such as nearest neighbor methods or SVMs. They can claim that they reason by similarity (to the past experience, similar episodes in the past), by making analogies, and that more than anything reflects the nature of our thoughts, as backed by evidence from psychology. Probabilistic programming and Bayesian learning methods can also reference evidence from cognitive psychology experiments in human infants and young children that their approach is brain-like. And finally, logic-based AI methods can reference our deliberate analytical thinking and emphasize similarity and universality in the way symbols are manipulated to make logical deductions.

Being aware of the massive potential upside of the technology, most of us strive to create truly intelligent systems. Unfortunately, the stronger this desire, the greater will be the confirmation bias to overcome. The confirmation bias on the belief that what we are currently building is intelligent or similar enough to the brain's intelligence. This bias is of course unhealthy because it downplays the shortcomings. It causes us to pick and choose cognitive or neuroscientific findings to reinforce what we hope to be true.

Another blind spot could come from the fact that modern computational neuroscience uses many of the same assumptions, abstractions, and methods that deep neural networks in AI do. That's understandable simply because reliably successful methods in neural networks are not easy to come by, and once they are found to be robust by the whole community over repeated experiments, they find their way into many other research areas including computational and cognitive neuroscience models of the brain. Thereby inducing a certain level of consistency between models of the brain and AI which may not be telling us the whole story.

Having said that, finding some unity and integration between deep learning and current neuroscience can be quite a promising avenue of investigation, and justifiably many researchers are excited by it.[162] However, as we argued before for the existence of many levels of emergent phenomena in the brain and the quality of understanding required to make claims about the phenomenon of intelligence, the following remains true:

It is extremely safe to assume that what we don't know about the brain far surpasses what we do!

Putting it all together we may conclude that it is best to avoid any assumption of a direct relationship between deep learning AI and human intelligence. Of course the semantics of "a direct relationship" is highly subjective. But we are just trying to make a choice here to see what kinds of statements and

---

[162] A. Marblestone, et al. "Toward an integration of deep learning and neuroscience." Frontiers in computational neuroscience 10 (2016): 94.

sentiments can help us most to overcome our confirmation biases and expose our blind spots, what sounds more intellectually honest, and what is more constructive for making progress. If we're truly after ultimate principles of intelligence, it's perhaps best not to put the "intelligence" label on what we already know how to do or build. Perhaps it's best to view intelligence as a missing manuon!

Fundamental Intelligence Hypothesis

So far we have discussed AI as just a label, setting aside questioning the meaning of the phrase. Finally, we get to change course, and open the door to use the word *intelligence* seriously. As we said at the beginning of the chapter, there are multiple intelligence-related topics that all overlap, but we gave the name *fundamental intelligence* to the ultimate topic of intelligence as a natural phenomenon, the true topic of the missing manuon of intelligence.

The thesis we put forth in chapter one was as follows: "to fully understand the brain's intelligence, we must go beyond the brain, abstracting away the differences between different examples of intelligence." We put forth the idea of an "intelligence science" where theories and principles of a fundamental intelligence are sought. One reason to give this concept of intelligence the title "fundamental", is simply due to the unifying foundation that such a concept could provide. Unifying in the sense that the study of intelligence would gain independence from any specific biological or engineering medium.

Another reason for the title fundamental can be much deeper and perhaps require facing the *fundamental bias* of our brain (discussed in chapter 3). Thus we present it as a hypothesis. The hypothesis is that intelligence is not only an emergent phenomenon in the physics of our everyday world but also a deeper and more fundamental fact about nature beyond what's relevant to human life. We are not yet equipped to fully discuss the meaning and ramifications of such a position as it requires lots of discussions about the nature of nature and reality, physics, and mathematics. We can, however, inspire the idea and give a brief rationale and motivation for it.

The concept of information has been increasingly spreading its roots deeper and deeper within theoretical and fundamental physics over recent decades starting perhaps with the phrase "it from bit" established by the legendary physicist John Wheeler[163]. Here's a quote from his 1989 essay on

---

[163] He is widely considered the most successful PhD advisor of all time with multiple Nobel laureates among his students including Richard Feynman. I was so fortunate to have one of his students, John Kaluder, as my own advisor in Grad school shortly before he retired.

the topic which sums it all up: "Information may not be just what we 'learn' about the world. It may be what 'makes' the world." What has been developed since then with fundamental physics, has only intensified the role of information. If we are going to reserve a place for "information" as a fundamental aspect of reality, then it's hard to argue how the flip side of that equation, namely "information processing", shouldn't also assume a fundamental role in nature. That is, it may not be a stretch too far to assume a fundamental role for intelligence as well or that there is a concept such as fundamental intelligence to be developed. As speculative as that may be, we'll take a stab at defining such a concept in the next volume.

Of course, any notion of intelligence at the heart of fundamental physics is currently an outlandish idea, if not outrageous. That is because there is not yet any place for a notion of an agent to which an intelligence could be assigned. And that boxes such work within philosophy of nature at this point, or perhaps philosophy of the origins of special-ness and simplicity of the laws of physics and so on. But that is not the only place where we can seek fundamental intelligence.

We'll have to talk more about the very word "fundamental" later, again in volume II. But for now, we must be aware that it could also refer to aspects of some emergent phenomena. In particular for us, an emergent phenomenon of intelligence as in the human mind. It deserves the label fundamental, because of the universality of what it represents.

In the absence of any science of intelligence, we should adopt notions of intelligence that are as universal as possible. That means attempting at frameworks that can explain emergent intelligence phenomena in both humans and machines by the same physics of intelligence. As we argued earlier the similarities between deep networks and the brain are due to some facts of nature, rather than a "direct" relationship. That too signals the existence of a common underlying physics of intelligence. Such physics can be studied as the real cause for the similarities and can establish some indirect relationship between AI and human intelligence. This extends to whatever machine intelligence the future may bring and other examples of intelligence.

That shouldn't be a surprise, given that all kinds of information that the brain or an AI has to process and act upon, are ultimately produced by physical systems. And from physics, we know how special (non-arbitrary) the generation process of all that data is. Those universal processes must be inducing many common features for any information processing or intelligent system. So far, "depth" is what we understand the most, as a common feature. In particular, when it comes to explaining why neural networks need to be "deep", people have already pointed out physics as the root cause.[164]

The hypothesis of fundamental intelligence is a bit more than some higher-level computational theory of information processing. It makes statements about nature and places intelligence within physics and mathematics. It legitimizes questions about the fundamental nature of intelligence.

First Principle Approaches to Intelligence

The idea that there should be higher-level computational theories for intelligent systems is an old one. Theories that would be placed at the first level of analysis of information processing systems according to Marr's 1982 categorization[165] (he considered two other levels of analysis: representation in an algorithm, and then the hardware implementation level). The idea is quite appealing simply because, using such a theory, we can build many different intelligent systems, implement them differently, on different mediums, and with different hardware. We'd be free to do whatever that turns out to be cheaper, faster, and better. That's what everyone's been hoping for.

However, there hasn't been any disciplined plan to get there. There are of course scattered efforts with the hope for common patterns to naturally emerge, patterns that we can then leverage to build a higher-level computational theory. The two main thrusts of such efforts have been one in AI and the other in neuro- and cognitive-science. As for AI, the approach has been by building systems through trial and error to perform well on tasks that we consider to require intelligence (something that keeps changing over time). In cognitive science, it has been by studying the brain with increasingly more powerful imaging technologies or by evermore sophisticated computational models in computational neuroscience. As promising as such reverse-engineering efforts may be, all such efforts are bottom-up approaches. What's missing is a top-down effort.

A top-down approach of course would start with speculations, perhaps in the form of axioms that are not provable, and for that reason, most researchers don't find it a viable path for making progress. It may sound like firing shots in the dark. We know there should be some universal principles of intelligence around. We do know that there is a giant target out there but we just don't know how far we are from it. That's exactly where top-down efforts could shed some light on. That

---

[164] H. Lin, M. Tegmark, and D. Rolnick, "Why does deep and cheap learning work so well?." Journal of Statistical Physics 168.6 (2017): 1223-1247.

[165] D. Marr, "Vision: A Computational Approach", San Francisco, Freeman & Co. (1982)

is, we could try to start with some reasonable hypotheses of intelligence and architecting theories or systems accordingly.

In trying to hook up many functions and AI agents to one another in solving the future problems mentioned in chapter 8, it's not too hard to see that the architecture of the solution must conform to powerful design principles. I call such a design principle, SUM, S-U-M, an abbreviation of three sub-principles of Scalability, Universality, and Modularity. Broadly speaking, they are the following[166]

- Scalability means that the same process (methodology) has to be applicable to small as well as large-size problems.
- Universality means that the same process (methodology) has to be applicable to various kinds of domains and problems.
- Modularity means that the same process (methodology) has to be applicable to various kinds of functionalities and subproblems of a problem and it should be able to function in pieces independently, and not just as a whole.

The name SUM beyond the abbreviation represents the addition of the three elements for a new whole to emerge, like a sum total. For that to be really the case, as in giving rise to some intelligence, the right kinds and combinations of these sub-principles must fit together. Our definitions above are not the only kinds. Even within the definitions above, there is a lot of room left for subjective interpretation or variations of measurement.

Different notions and instantiations of these underlying principles of SUM can give us different systems to be used for different purposes. For instance, the definitions we gave above can capture the essentials for handling the interactions among prediction machines, incorporate the more general decision-making-view of intelligent networks, and facilitate decomposition learning as explained in the deep learning section (take deep feedforward neural networks as a special case).

Modularity plus scalability means we can hook many different pieces together to make a functioning network and also hook networks together to make bigger networks. It also means there should exist some elementary units of processing. We happen to call them neurons. It also means we can do things like bringing several different networks together and pass them as input to another

---

[166] Definitions are selected from: A. Magin, "Integrated Intelligence Systems and Processes" 16/365400, United States Patent Office.

network that mixes them. That is in fact one main way multi-task learning gets architected. These may sound trivial but, as design principles, they can be incredibly constraining. Universality plus scalability means different neurons should work the same way independent of where they are placed in the network, they should have the same principle. Computational scalability and statistical scalability mean we should have both width and depth as we explained in the deep learning section.

Just because historically we arrived at deep neural networks through decades of playing with perceptrons, decades of playing with backprop, and thinking so much about how the brain does it, doesn't mean we couldn't have thought "deeper" much earlier without any experimental validation. Well, perhaps the pioneers who did stick to their guns and waited out winters till there is a lot more compute power and data, did so for these very reasons. We can't quite tell. What we can tell for sure is that we should be able to arrive at many promising systems through first-principles such as SUM without any knowledge of biology, the existence of the human brain, or trying to improve upon existing architectures, *provided that* we could let go of only working on things that we can get them to work right away (meaning we know for sure we'll get something out of)!

However, the challenge remains with more precise definitions of any such first-principle. As for SUM, that translates to asking what are the more precise and appropriate notions of scalability, universality, and modularity to be combined. There may be many more dimensions and degrees relative to which a truly intelligent system must stay SUM-able (scalable, universal, and modular). For instance, computational and statistical scalability in deep learning are only defined and measured relative to shallow architectures, and local estimators.[167] Higher standards for intelligence may necessitate much stricter scalability requirements. So far, researchers have been very clever in leveraging these principles informally, relying on their intuitions and good engineering instincts. Though a more formal investigation in this direction can be of illuminating value.

Now, if a principle like SUM is to be a principle of a higher-level theory of intelligence, it should also be applicable to the study of human intelligence. The brain does, in fact, have powerful design features that could be summed up in a SUM principle. The co-existence of specialized brain regions, along neuroplasticity, homogeneity especially across the neocortex, and the appearance of identical principles at work among the various networks and functions, altogether scream scalability, universality, and modularity.

---

[167] Y. Bengio, and Y. LeCun. "Scaling learning algorithms towards AI." Large-scale kernel machines 34.5 (2007): 1-41.

Chapter 10

Is AI in its Behaviorism Phase?

L et's now set aside attempts to give rigorous meaning to the word intelligence and return to AI as just a label. Whether it is just fake intelligence or not, just imitating intelligent behavior or not, we already have very powerful techniques in the toolbox labeled AI. As we said we can turn many problems into a single function approximation problem, or some collection of function approximation problems to approximate things that go beyond functions (see appendix 2). We can not only hook multiple networks to each other in clever ways, and train them in creative ways but also assist the networks with external or background knowledge and even "baking in" knowledge from the physics of the problem into the AI system in myriad ways.[168] All that without even talking about the power of hybridization which we have also discussed. The possibilities seem endless and quite promising.

All that is still without mentioning the sheer power of just a lot more data and computing power. Simply making a network outrageously large and training it with massively more data is also bound to give fun and exciting results somewhere. That is only an option for very few organizations with relatively unlimited budgets. You know, OpenAI or Google's Deepmind's of the world, who are living up to the hype they contributed to creating.

Great, given that there is a lot of good we can do with the methods we already got, we should ensure a healthy and steady flow of funds to further AI research. News of exciting results do just that. On the other hand, all that hype and excitement make it really hard to reflect clearly on how

---

[168] See the active DARPA program on "Physics of Artificial Intelligence (PAI).

much blind trial and error is involved here to make any progress, and on whether the progress is towards any real intelligence or, for lack of a sharper term here, fake intelligence.

Previously we argued that inspirations from biology and similarities between deep networks and the brain do not constitute any direct relationship between the two. In mainstream AI research, however, this is not of any relevance. The only similarity between human intelligence and AI that is of any real concern is the similarity in the set of capabilities and performances. We want to be able to do what humans do and more. The goal as often said is just to build useful systems. As a result, we truly care about *what* a system can do and its overall performance in tasks we choose at the time, but not really about *how* it does it. The how is of course discussed but no serious judgment is passed on it, as long as it achieves certain results.

Recall from chapter 3 that behaviorism established itself as the dominant school of psychology in the first half of the last century, by rejecting any theory or prediction that couldn't be measured in the external behavior. Later radical behaviorism claimed there should be a way to measure any relevant cognitive phenomena like various mental states in some external behavior too. Effectively, rejecting anything else as non-scientific just because they couldn't be measured at the time.

Physics has not been a stranger to bad philosophies either. Dismissing ideas that involve unmeasurable constituents as non-scientific is what *positivism* encourages. Positivists persist to only think in terms of things that we know can be measured. This philosophy not only has done damage in the past but likely to cause more in the future too. Take statistical mechanics, a giant milestone for 19th-century physics, as an example. Not only it wasn't considered a milestone at the time, it was also rejected by many who subscribed as positivists, simply because they couldn't see atoms or molecules at the time. They'd argue we should have focused only on the measurables like heat and energy. Combine that with *instrumentalistic* insistence to value things (theories in this case) only by what they can explain or do right away, and you'd have even less reasons to take statistical mechanics seriously. Again, simply because you are far away from seeing its power in action.

In fact, every field seems to initially get dominated by an instrumentalistic phase until it matures up a bit just to reject such philosophies as too myopic. It happened to physics, it happened to psychology, can it happen to AI?

Wait, what does that even mean? Well, just like psychology used to reject talking about stuff in the brain we can't see or measure in behavior as non-scientific, AI research today also boxes itself narrowly in beating benchmarks defined around the overall behavior of AI systems.

Perhaps, AI is in its behaviorism phase just like psychology was 70 years ago!

The current measures (including the proposed ones) are all defined in the space of "what is done", not in the space of "how it's done", how it would be done, or the space of methods, if you will. It is not yet in the space of understanding and the ability to do without actually doing. The reason we need measures on the ability to do something without actually doing, is the same reason that prompts us to distinguish a few intelligent acts from an intelligent person. We tend to give the full intelligence credit to a person, only when we believe they could accomplish many complex tasks that only the future will make known. We are keen to infer their overall ability to accomplish new goals, adapt to new conditions, or solve new problems. We attribute that ability to a behind-the-scenes intelligence feature. What's noteworthy is that, in measuring that feature, we know that HOW they do things matter much more than WHAT they do. We know what-they-do is temporary, while how-they-do-it stays with them much longer, it is indicative of who they are.

It is true that no one can articulate well or describe the algorithm by which we pass such judgments on people. For that reason, it is likely that our personal judgments in this matter are often quite flawed. However, just because we don't know how to measure something or measure it accurately yet, doesn't mean it can never be measured or it is non-scientific to talk about. That's the stance that behaviorism took and we shouldn't!

Of course currently, in AI systems, we have no clear clue on what such measures on how the system does something could even mean, let alone actually measuring it. We only understand tasks that are independent of methods and various performance measures on those tasks. We can see the AI making very silly mistakes but we have no idea how to quantify silliness. We can see there is more to learn from the examples or data we provide but we don't know how to measure how much more the systems could have extracted from the data except through pre-designed tasks one at a time.

I am tempted to call all the measures we currently have, to build and judge AI systems by, "active measures", in contrast to "passive measures" that don't necessarily require acting on any specific task except for the tasks of playing with the data to build confidence that you would do well on many tasks without even testing yourself. The question is how to quantify that confidence and that is something we haven't even considered investing in yet. That's exactly what being in a behaviorist phase means!

Being in a behaviorist phase is not necessarily a bad thing. If anything, just like behaviorism contributed massively to getting us past "depth psychology" and the Freudian school, opening doors to a much more scientific pursuit, the mainstream research in AI is also moving us forward and opening doors to new opportunities. One of these opportunities is for us to recognize that we may get stuck in the AI's behaviorist phase, and we can avoid such a bad trajectory.

Being stuck in behaviorist AI doesn't mean there wouldn't be any progress, it just means the progress is not built on the right foundations. When things are not on the right foundations, they can go only so far. Not unlike the case for building construction. An example of what could force us to stop and rethink foundations was finding ourselves in new regimes of operation in real-world applications, like those we discussed in the "future scale issues" section.

As it is often entirely possible in science, we may not get to know what the right questions or the right "passive measures" are until we are already close to answers or have acquired the ability to perform well on would-be good measures. Without a proactive approach, what typically happens is that we come across conundrums that we don't even have the language to talk rigorously about. We get forced to rethink and a period of chaotic debates precedes eventual resolution and robust progress thereafter. Then as always, our hindsight would be 20/20. The 64-million-dollar question is: Can we be wiser this time and develop our foresight to get any closer to our would-be hindsight?

It should be obvious that the first step towards any breakthrough is believing that, when it comes to intelligence, we don't exactly know what we are doing. A critic may say "but I don't see how we could do anything else right now that would be science and we aren't already doing. If you can't measure the performance against tasks the community has established, how is it science?" That mindset is partially due to the bad philosophies of science that portray a real and objective separation between science and philosophy. In reality, there is no such separation. Science is a human endeavor like any other and it's built on a mixture of philosophies. That is exactly why we need to talk about the philosophy of science in the next volume.

At any rate, it shouldn't matter whether we call it science or not. What matters is that it's the next step, to figure out the right philosophies, foundations, and measures that enable more efficient progress, again as judged by would-be hindsight. Working on the foundations is exactly what sheds light on what could become measurable and then we could get creative on how to measure them. That's part of the attempt in the second volume.

Lastly, let's acknowledge that many people reject the premises and the message of this section. They must believe that we don't need to bother with anything except to continue to focus on building better models upon each other's work, by trial and error, and by using our "best" arguments and guesses. Their premise is that we should continue to use ***AI as just a label***, and at some point, the right meanings and principles of intelligence will naturally pop up.

My valued reader, having read this book, what's your take?

Chapter 11

A Fictional Viaduct to Volume II

This chapter is only to serve as a bridge to volume II. To take a break and to stretch our minds a little, we are going to jump into fiction now!

Adam Bani's New Challenge

You're Adam G. Bani. You're a genius. You're at the right time at the right place too. You're lucky. You work for an awesome defense contractor who is about to get rich. Your boss walks into your office, shuts the door behind, and goes "this is a bit of an emergency and from this point on, anything I show and tell you is top-secret classified". She opens a thick folder, searches through 100s of pages full of calculations, figures, and geometric-looking patterns, and pulls out the following report for you:

Pre-meeting

About 20 months ago, the "space force" began to record anomalous signals coming from what appeared to be a little dust cloud near Neptune only about a million miles away from it. A dust cloud that seemed to be on a spiral trajectory that would eventually fall right on our Sun. A highly competent and diverse team of experts was assigned to investigate the matter further. Soon it became clear that perhaps it's not us trying to deal with some dust cloud, rather *it* was trying to deal with us! The signals were not sporadic, they were constant and they kept changing only once every four hours. Each time a new pattern of photons was received. The approximately four hour lapse between distinct patterns seemed to be carefully tuned to be equal to the time light takes to travel the distance from them to us. At this point, all bets were on something at least intelligently-designed approaching us, that at a minimum does some form of physics calculations and performs computations to execute on some intention, based on its calculations. If this was intentional, and these signals were in fact being generated for us on earth, the patterns had to start changing faster as they got closer to us. And it did, their pattern-change frequency was exactly proportional to the change in distance. If they were not coming for us, the messages were definitely meant for us. It wasn't clear at all what they wanted from us. Their messages had to be embedded in these varying patterns of frequency, intensity, and geometric shapes of the luminous part of the dust cloud. Meanwhile, more precise calculations revealed that their spiral path was not a classical one for its apparent mass and typical make-up of cosmic dust clouds. Something was very novel about either what it was made of and its distribution or how it was moving in and interacting with space! Simulations based on collected data and assuming the same behavior pattern would continue, it looked like they would end up in our neighbor's yard, an orbit around Mars!

We expanded the team and gave it top priority. After a month of effort by our devoted mathematicians, physicists, and computer scientists, we were able to decode the signals. They turned out to be a bunch of different cartoonish paintings and gestures with labels and captions which were expressed in terms of constants of nature like the speed of light, Planck's constant and so on that we call *natural units* in our world of physics. Finding some common understanding was exciting but also frightening. There was a lot in these messages we couldn't understand but some of them could trivially be chained together to form a cartoon with a clear story. A story that starts by portraying the dust cloud on the path that we had witnessed part of it already. Interesting was the portrayal of what would or should happen next. Their story seemed to play out by us leaving Earth in what would be about half a year from the time of decoding their messages to meet them on an orbit around Mars in about an extra 6 months from then! Was that an invitation to literally meet up?

Regardless, it was absolutely remarkable, their calculation of Earth's and Mars' orbit around the Sun and our own trajectories to meet up on time in a particular location in space, a whole Earth year ahead. As mind-blowing as it was, at least this story seemed to be agreeing with our own calculations of where they were headed. The cartoon depicted some form of a face-to-face dialogue between us at the meeting by exchanging light waveforms. It also appeared that we're showing them recordings of what we humans exchange with each other when interacting with one another. It showed very basic forms of exchange. It had to be about information exchange. It was showing them learning our language by unclear means.

We were on high alert, extremely suspicious. Yet we couldn't help but wonder that if they wanted to harm us, why would they give us so much room and time in advance to prepare. Why would they even message us in the first place instead of "driving straight to earth with their headlights off"? The dust cloud didn't seem to be massive at all compared to planetary masses, it was just spread wide, looking like a cloud. So, there was no threat of them being able to cause major damage by any dumb collision. On the other hand, maybe they could increase their speed to relativistic values and thereby increase their mass!

Their speed was observed to be about 0.1% of the speed of light. Our speed would be about 20 times less than theirs going to Mars and relative to Mars. They not only seemed to know that but they had taken it all into account in their calculations which had given us exactly 6 months to understand the messages, prepare for a trip, and 6 months to get there. Surprisingly, the start for our trip was carefully chosen at a time that we'd be traveling a relatively short path to Mars, which could save us about 2 months in Earth time. Why would they spend energy and compute cycles or whatever it is that they are doing, to try to make anything more convenient for us?

Would you trust such an alien invitation for a blind date? The more we realized how intelligent this dust cloud must be, the more of a risk it seemed to ignore their messages. They wanted a meeting and we prepared and sent a devoted team of astronauts to give them just that. We had to follow their scheduled plan per the cartoon they "mailed" us. We gathered all sorts of investigatory technologies for our meeting from devices to intercept and record all their messages and activities to probing their internal physics, should they let us. But we also packed up cutting-edge weapons, the latest and greatest in our "space force". We didn't forget that we had to find a way to establish a common language per their hint. We took about a thousand movies and TV series with captions in multiple media forms ranging from spinning hard disks to invisible-ray disks (the technology that was coming to replace Blu-ray and never hit the consumer market, thanks to the digital Cloud and streaming services like Netflix). We also took copies of millions of books, thousands of textbooks, an archive of many things on the internet, including our social media and of course a complete mirror of our arXiv.org library!

Meeting

It took us about 6 months to get there as projected. We are ready and instructed to use our space weapons if we sense any danger except we weren't even sure what would work on this cloud if any. As we were approaching, the cloud illuminated itself in many spots all blinking synchronously. The light coming from it was monochromatic, a single frequency. But the choice of frequency was interesting, about 500 THz, i.e., the kind of yellow that our Sun radiates most intensely in its spectrum. That way they were making sure it'd be a familiar frequency for us and therefore maximizing the likelihood that we'd detect it. That looked like a super friendly and welcoming act by them.

We get to their orbit right behind them, just a few miles away, which means we're up their nose given typical length scales in our solar system. The cloud seemed to be miles long in every direction. As we got situated, the cloud seemed to be deforming the part facing us into a giant movie theater screen. It started showing a version of our own spacecraft. The colors were all fuzzy, but it was clear to be an attempt to portray us. Looked like they had instantly analyzed the patterns of light diffracted from us. As we are wondering why they are putting up a strange mirror in front of us, the image started moving, it was showing us throwing them a bunch of things, boxes of things indeed. The boxes were then shown to open up and emerged from it the same pictures they had drawn for us in those messages we got from them back on Earth. In particular, the ones we interpreted as referring to our language and knowledge!

Next, they showed a cartoon version of themselves (pretty much a dust cloud) receiving the boxes and probing them. So, it was clear how they wanted to start. They kept repeating this from different angles, they were showing how the cloud could deform and adjust itself to receive, stabilize, manipulate various things that could be thrown at it. Whether that was a hint for us to send them whatever we want and not worry about size, form or interpretability, wasn't clear. We decided to shoot off all those relevant educational content to them anyway. Along with that, we sent various devices that could read them in case they could use it, laptops, various media players, etc. Most important of all we dared to send along a number of spying devices that could store information locally and also send it back to us remotely. They were a whole suite of tools and layers to capture not only the patterns of the read access to any of the content we sent in full detail (order and depth of access at any granularity with a hierarchy of timestamps and so on) but also any of their internal communications and computations that we could record with our most comprehensive hybrid sensors. All to gain insight as to how this form of matter or creature operates.

It was magical to watch the boxes being absorbed by the cloud. They seemed to be distributing the content across the cloud and acting on them in parallel. It wasn't clear we were dealing with a single entity, i.e., the dust cloud or many entities in well-calculated dynamic coordination. Certainly, there was no obvious central unit that we could detect. We thought maybe we're dealing with some sort of a floating ultra-harmonious "society of minds". Turned out they were not going to let us wonder for too long.

In a matter of minutes, it started to reform the same display screen for us. As it was forming, some fuzzy mixture of snapshots from various movie scenes we sent them was popping on and off on the screen. That all faded away once the screen was flattened out and suddenly went pitch black, no visible sign of the cloud now. Was that a strange rehearsal of some sort? As we were wondering, the following words all together were illuminated across the patch in space where it has just disappeared into:

"We're glad we got a chance to meet you. We suspected you'd be here somewhat struggling to figure things out, as evident by the form of your language. You call that English, right? And you call yourselves HUMAN, right? Or UMAN if you're from New Jersey, USA perhaps? Did we get that right?"

We were rendered speechless. No one had expected THAT. We thought we had lost our minds. So much so that some of us started to punch ourselves in the throat. We even thought to test for traces of Chinese Fentanyl on our last course. Thinking: how else can we be so high right now? Or are we? Either way, there is no reason not to lead on - we're thinking. Too long was passed in silence so they changed their displayed text. Next screen read:

"Thanks for coming on such short notice, we'll tell you why we bothered you to come all this way to meet us. But we need one more thing to conduct our meeting more efficiently. A way for you to communicate your replies, questions, or comments. That way we can start to get to know a bit more about each other"

The screen went dark again, it looked like they were waiting for some form of response or action from us, perhaps any. We sent off a digital board roughly the size of an NBA player; 2 meters by ½ meters, with large pixel sizes such that it could remotely display max 140- or 280-character long messages on it. Now we could send them a "tweet" and reply to theirs, except theirs were being printed with a font size 1000x ours and perhaps with no limits on the number of characters. As soon as they got our digital board and had it locked in a position, their screen went on again:

"This is very helpful, thanks. Based on the materials you sent us earlier, we now have a fairly well, and consistent understanding of you, your biology, and how it's different from our own ancestors. Your math and physics are a lot more similar to that of our ancestors than your biology. Your language is at a very preliminary phase though and is struggling with what our ancestors seem to have struggled with too"

Our curious captain jumped in with exactly 140 characters:

"We do have other languages besides English, that can describe subjective matters and emotions tons better and richer, would that be helpful?"

Cloud replied:

"That's not exactly what we mean. We got some of those other languages too among various things you sent. Thanks for including a diverse set of languages but seems like all of them are only one-dimensional sequences of symbols, which is a very poor and limited representation, with very low knowledge transmission capacity. However, we did notice you break those patterns in what you call "art", in your two-dimensional messages (e.g., images or paintings) or three-dimensional messages (e.g., videos). But you already know that there are novels you can't fully put into a picture and vice versa. We highly suspect that you must sometimes be noticing a richer language being spoken within your mind (the word that you seem to be hiding most of your lack of knowledge and misunderstandings in it). You want to describe reality in terms of rigid and fully separated symbols or shapes, put in a line one after the other. Seems like you haven't changed your attitude and desires much since Plato's Forms."

We didn't expect that and got more puzzled as to why the heck do they care and why would they talk down to us like that right off the bat. We tweeted:

"Okay, good to know. What do you want from us?"

Their next screen came on:

"First and foremost, we are by no means here to cause you any harm. If anything, we feel some nostalgia towards you and we know you cannot quite understand that and that's OK, you will understand our win-win plan that we will get to shortly. Let us give you some background first. You seem to call entities like us and yourself as examples of life and intelligence. Let's go with that. Our ancestors were also a form of intelligent life but a few billion Earth years ago. Their biology was not quite like yours but still was based on similar chemistry. For instance, their lungs were also made of atoms and molecules (very stable forms of matter in this universe) except theirs were made to breathe in hydrogen while yours attempts at getting oxygen. Anyways, our elementary versions were given birth in giant laboratories. Fast forward billions of years, here we are the advanced versions bypassing and countering this brutal accelerated cosmic expansion, that is driving us all apart for eternity."

Before their next screen lights up we had to jump in again:

"Sorry for the interruption but not so fast please. We got it about your ancestors but how do you operate? How do you have life? Do you breathe?"

Cloud replied:

"Our 'lungs' are made of a tiny patch of what you call space-time. It's not just our lungs, it's our "hearts, stomach, legs, and feet" too, that's where our energy comes from and also how we move. "particles" pop in and out of existence in our lungs and make this happen. Most of our "stool" is, how should we say, a very particular kind of "dark matter" (the label you are using to describe many things that you currently seem to think is just one thing), if we may leave it at that. There is nothing static about us. We have several stationary modes and operate within their bounds. That is our biology if you will. If these stationary processes that give us our 'life' fall out of their stable regions, we can't recover. We turn into some dumb blob of matter, i.e., we die. We are made of a hierarchy of many bound states, with dynamic bindings between what you call "empty space-time" and sub-space-time. We use space-time itself to propulse in it."

We were trying to digest that when the next screen came on right away:

"Please excuse us if we don't have an easy or useful way to explain things further to you. They may not bear much significance for you either at this point. Here's what we want you to know. We know you guys conduct a lot of experiments; use gravitational wave detectors or high energies of accelerated particles to probe very small distances or times. We are here to set up a giant laboratory on Mars that could benefit you in the future a lot and we are thinking you could very well help to maintain it for us. We are confident that once you advance sufficiently to be able to use it properly, you most likely have also developed goals that are overlapping with some of ours. The accelerated expansion of space in our universe is going to be causing a lonely future for our galaxy, driving all other galaxies outside of its entire visible universe. Currently, you seem convinced that what the future holds is most likely nothing in space but the ever-expanding space itself and vapors of black holes. We understand how you have arrived at that conclusion. The good news is that we think differently. It's going to be simple but not boring and depressing like that at all."

We jump in once again:

"That's some deep stuff, if you know so much, can you just tell us why we are here? Why do we exist and what should we do with our existence? Do you know?"

Cloud responds:

"Don't you see the massive 'phase space' of the universe? How can almost all this giant phase space of possibilities exist so as to just never be realized? Don't you feel that's odd? If so, it's because it is. Certain things are too hard to explain in your language such as brute facts: things that just are! Just like being is! The universe must realize its existence for its being to actually be. It's better to ask about our "role" in the universe than our individual purposes! Our minds are made of "time", and for "time" (the emergent macro time)! That is going to be our role letting the universe experience more of itself. Our role is not to demand meaning from our existence but give meaning to existence by letting it realize itself: to be. That means our immediate purpose is to help "this macro time" to go on! If you think that is about "increasing entropy" as you call it, yes, you are getting close but it's more like accelerating it with ever more sophistication.

Their next post came on immediately:

"Let's not get ahead of ourselves though. We need to procreate too just like you do, but with us, it's mainly to proliferate ourselves even though we can't quite catch up to "space proliferating itself". Our babies will be born in these labs based on our recipes. A bunch of us entangle together and create new patches of space in which we gain some form of locality, i.e., "access to each other". We are never truly alone in that sense and the "sky's the limit" for the kinds of things we can do.

It's not like we are without any vulnerability. We explained to you we can die too. That's why we must be careful. If all our processes fall out of their control bound, we all die. Imagine if all DNA molecules on Earth were wiped out. All life would be irrecoverably wiped out. We don't want that to happen to you nor a similar version of it to us. We know that you are not a threat to us at all, at least not anytime soon. You may find this hard to believe that the more intelligent you get, the more trust we can place in you for a mutually beneficial partnership. We would like you to advance to a point to want to share our goals."

We asked:

"That's quite interesting and suspiciously nice of you. Advance how? What partnership? when?"

They replied:

"We can also help you create a human-habitable environment here. We'd like you to be part of the facilities that we'll set up here. And it's OK to be suspicious of us. We have our concerns about you too but it's about lack of trust in your competence, not your intentions. Unfortunately, we can't yet leave this planet (Mars) to you right now. Your current population is far below what your own planet can handle, yet you seem to be really struggling to take care of its environment. Please excuse our candor, but there seems to be too much stupidity and suffering going on back at your planet. If you can't get rid of it there, you won't be able to get rid of it here. Giving you access to new technologies is not going to fundamentally change you and what drives you. You'd just use more advanced tools of tomorrow to pursue pretty much the same desires as yesterday. If it's war, you use more advanced tools to win the war. If it's peace, you use it to create too much inequality to cause a war. Same underlying stories and motives, different day, different context. And by you, we mean 'the collective you'. You can look at yourselves as a bunch of individuals or as one collective entity. Both ways of talking about the "reality of you" are valid. However, the emergent stupidity is easier to observe in one than in the other. What should be surprising to you is that several pieces of your own literature explain these fairly well already. For instance, we found this verse of a poem called 'Bani Adam' by the Persian poet Saadi Shirazi, quite concise in that regard:

"Human beings are all parts of one body",

where he states that the pain of one part will spill over to the others. And that was back in 1258 AD. It appears your civilizations have forgotten about the very basic foundations of your collective being. You see us as one, and yet you read that we refer to ourselves as "we". We are one and we are many. That's our fundamental state.

Our analysis of your micro and macro histories tells us that the majority of humans, if not all, thought either too little or too fast in many cases. As it stands now, you'd be doing the same on Mars. Having said that, we are here because we think you have great potential. Otherwise, we didn't have to be here, we could set up our lab anywhere else, not that we haven't been doing that. Our hope is that you could use this as an opportunity to improve yourselves and embrace a very exciting future. That means for you to try to get your acts together back at Earth and get a lot more intelligent."

That was offensive to us. We felt we could show off a bit by:

"We recognize that of course and we are working on that but we have a lot of limitations, especially the biological ones. However, we are very hopeful that one technology, in particular, we have been developing could help us break away from our past situation. That is what we call "artificial intelligence" (AI). That's how we are hoping to become more intelligent and do a lot more good than bad although it can go in many different ways. Like we said it's a work in progress, but we already have a lot of promising results that make us feel we could get to your level or a level that could amend your needs and be able to earn your trust."

They posted their comment next:

"We noticed that vibrant field of research and technology. However, you're moving quite inefficiently. Your culture and institutions force so many brilliant people to publish so much garbage. You force open-minded people to act so narrow-mindedly. As a result, your system rewards fancy and fashionable complications rather than elegant simplifications. And true progress is within the latter. If you want our friendly advice you may want to start from scratch on that one, you'd move forward a lot faster."

We asked: "If you know so much, why don't you just tell us the answer in simple terms, what we are doing wrong, what the solution is? Tell us the principles and recipes to create better AI."

As if we hit a nerve with that question, they instantly lit up a few back to back posts on their space-made theater:

"Well, that's not the right question to ask nor a meaningful one. For instance, defining 'Better AI' requires philosophies you don't seem to be close to forming. And we wouldn't just plainly give you 'principles and recipes'. Let's take a look at your recent past. Didn't your own physicists give you a bunch of principles and recipes, which you then turned into nuclear weapons and actually dropped it on yourselves? What are you doing about it now? Can you say that the nuclear arms race among yourselves is totally over? Those physicists were smart, right?

Throughout your history never a newly discovered fact about the world came with a theory that combines you and the knowledge of the new fact as intimately interacting parts of the same universe. When it comes to that, you seem as clueless as you have ever been. Your Isaac Newton said he could calculate these orbits, like the one we're on right now as well as those of the planets around us but he couldn't calculate "madness of men". How can your understanding of nature be so asymmetric or is that you think that you're not part of nature? You have always considered discovering a new fact and what to do with it as totally independent matters, and should be worked on by different people. Has that been the right choice not to work on the latter with equal rigor? Were you not smart enough to work on these problems or not humble enough to consider them as problems?

While most of you want to just pretend you know what you're doing, some of you admit it's a problem, but even then, not one that matters, because new facts will always be used by those who just want more power and want to control more resources. Let's do anything because if you don't do it, someone else will. right? If you don't gain more power someone else will. Have you noticed how ubiquitous this "logic of inevitability" is? You have so strongly convinced yourself of it for it to be behind almost all the history-shaping choices of your past. It has even dictated the fundamental structure of your societies.

As we pointed out before, you have made a lot of technological progress using the new facts you discover about the world. But YOU'RE STILL YOU if you know what we mean! You get your work done with fancier tools and faster. You travel faster, you live longer, your economic productivity is higher, etc. But do you know any better why you do this and what you want? Are people better and living more meaningful lives? Has endless philosophical debates over 'meaning' and the 'meaning of meaning' resulted in any real progress? Or is that the case that your dominant mental patterns and the emotions which move you, namely, greed and fear, are still the same forces shaping your history and your future as they've ever been. The same type of fear and greed are as always controlling your destiny more than anything else at the core of what you are and what you're doing. Or, is it that you think that's exactly what the universe planned and wants from you?

Greed of some of you turns ultimately into the fear of all of you! And the collective unity of you, is being fundamentally driven by fear. That's a choice, not the plan of the universe for you. Without you knowing, by that question, you are asking us to help you live in evermore greater fear. **You are asking for help where you don't need it, while you need help where you don't want it!** That is because you assume you really know what you're doing. Why do you want AI? Just because it helps you continue the same behavior as before more efficiently with more advanced tools? If we gave you any technology today so that you can do with it as you wish, it won't solve any of your existing problems, it will instead surface them more as well as create even greater problems for you. Technology without a corresponding "theory of you", will only magnify your problems. Don't you see enough of that already around you?! The greater the technology, the greater your understanding of you must be to have a complete theory. They go hand in hand, there exists a nice symmetry between them that you keep ignoring and want to keep breaking in your asymmetric and incomplete theories.

You have broken the symmetry and it has broken you!

This asymmetric growth of your knowledge about nature, where you and your role in it are mostly missing from the picture, makes knowing how to use something for "good", a very hard science for your current civilizations to crack. However, you have a fantastic opportunity right in front of you to change that, should you not miss it!

A great deal of your respect for yourselves comes from your intelligence. When you are developing something else intelligent, you get to reflect on yourself a lot more. Not only to draw inspiration from it but to figure out what you should do with it and how to have it co-exist and interact with your intelligence. You can use this opportunity to question who you are and what you should want. Creating something intelligent and designing an environment to use it beneficially, more than anything can help you learn about yourself, should you choose that route. It's a real chance to create a framework to formalize and tackle humanity's problems that have always existed. Of course, it can also go a different way, like any other technology in the past, where what it should be used for, stays outside of the rigorous set of equations. However, with AI it would be a greater missed opportunity, simply because it intimately relates to so many aspects of being a human, not to mention that many of your efforts directly study humans in order to replicate something similar in AI.

Like we said before, what we wish for you, is for a totality of you to become a lot more intelligent, for a majority of people to get a lot smarter and until then you'll still be you! And if you think that's never going to happen, we wouldn't be surprised. Because that is so typical of you humans, just because you don't know how to do something, you say it can't be done. In case you're wondering, you should also know that any intervention by us to help you on these matters wouldn't quite work, even worse, it could cause a lot of unintentional suffering for humans, you know how emotional humans can get. Any one of you astronauts who feel brave enough to put this to test, can go back home and suggest that we are coming over to help humanity get smarter instead of the machines you want to create. It's safe to assume that no amount of screaming "Don't shoot the messenger" is going to save you!

Having said all this, which you should take only as friendly advice, if you still want help on AI, we have nothing more to say than repeating ourselves that you should work on basic principles,

principles that you're far from in your current mindset. So, you may want to rethink the issue. And if you have no other questions for us now, we are going to send you back the materials you kindly provided."

Seeing no post from us, they sent all the boxes back including our spy devices, while displaying:

"It was a pleasure to meet you. We'll be in touch! :-) "

We got all the materials, and it took us another 6 months to come back from that 60 minutes of "face to face" meeting.

Post-Meeting

Before any rigorous analysis, there were already many different opinions on the table, with regards to the actual threats involved in this situation. Of course, we were not going to be naive and just believe their story of course. One thing that was strange was them not detecting or saying anything about our spy devices if they did detect it. We were wondering if they intercepted the inner workings of our devices and modified it such as to give us only the ideas they wanted us to take away. But we ruled that out quickly, as our sensors would simply not work consistently in that case.

We analyzed the hell out of all our recordings, patterns of access, and traces of computations by the cloud. Many AI and signal processing experts unanimously declared that the way they learned our language for instance was nothing like any of our methodologies. Their efficiency and ability to learn so much from so little so quickly was mind-blowing for everyone. Not only it wasn't like any of our systems, it wasn't similar to the way a typical human learns either. It wasn't like any of our humans nor machines. Whatever it was, it demonstrated quite an intelligent behavior. One that we certainly need to understand. Regardless of this cloud's true motives and what it says, the risks are enormous for Earth and our civilization. We may not agree on what the risks are exactly but we all agree that we must enhance our capabilities by quite a lot and as soon as possible.

This matter holds top priority and we, the company, are allocated a $100B budget to it, with a considerable portion to be spent on basic research. Based on our investigations, and the main body of this report, we want a massive reset in the mindset in this research. Therefore, a significant portion of the budget for basic research will be designated to bring fresh mindsets in order to rethink the field of AI.

You finish reading the report. You cannot determine your own feelings: whether you're more nervous or excited, whether it's more of an opportunity or a threat, etc. Perhaps you feel dizzy more than anything. You lift your head and see your boss staring at you waiting not so patiently for you to say something. So, she goes "there you go, this is the challenge from the space force. You can draw from an exclusive $4B budget that we got to rethink AI. Use the money to make anybody you want on your team to quit their current life tonight. You've got only 6 months, beyond which period, access to the fund may not get extended."

What Now?

Of course, this was all fiction, but let's be open minded and ask how can we best judge such fictions about AI? Could the story, when it comes to the intelligence part, ever materialize? Could this level of life/intelligence ever manifest in any way? Taking the challenge would be the way to answer this question! Good news is that we have started a bit on that journey already. In this volume, we tried to review history and uncover some assumptions and tried to clear the dust a bit, but we are yet to discuss many topics that could ground a fruitful definition of fundamental intelligence. That's what we'll pick up next and try to get out of our current thinking box!

Additional References

Chapter 1:

Hawkins, Jeff, and Sandra Blakeslee. On intelligence. Macmillan, 2004.

LeCun, Yann, Yoshua Bengio, and Geoffrey Hinton. "Deep learning." nature 521.7553 (2015): 436-444.

Mitchell, Melanie. "Why AI is harder than we think." arXiv preprint arXiv:2104.12871 (2021).

Lipton, Zachary C., and Jacob Steinhardt. "Troubling trends in machine learning scholarship." arXiv preprint arXiv:1807.03341 (2018).

Chapter 3:

https://plato.stanford.edu/entries/kant-spacetime/ — substantive revision Mon Oct 10, 2016

https://plato.stanford.edu/entries/kant-science/ — substantive revision Fri Jul 18, 2014

Jenkins, J. J. (1980). Can we have a fruitful cognitive psychology? *Nebraska Symposium on Motivation, 28,* 211–238.

Wilson, Robert Andrew, and Frank C. Keil, eds. The MIT encyclopedia of the cognitive sciences. MIT press, 2001.

Chapter 4:

H. Gardner, *Frames of Mind: The Theory of Multiple Intelligences.* Basic Books, 1983.

Bringsjord, Selmer and Naveen Sundar Govindarajulu, "Artificial Intelligence", The Stanford Encyclopedia of Philosophy (Summer 2020 Edition), Edward N. Zalta (ed.)

Stuart J. Russell and Peter Norvig. Artificial Intelligence: A Modern Approach. Prentice Hall, 1995.

Giaquinto, Marcus, 1987, "Review of The Rationality of Induction, D.C. Stove [1986]", Philosophy of Science, 54(4): 612–615.

Hume, David, "A Treatise of Human Nature, Oxford", Oxford University Press, 1739.

Franklin S., Graesser A. "Is It an agent, or just a program? A taxonomy for autonomous agents" Vol 1193 (1997), Springer, Berlin, Heidelberg.

Ariely, Dan. Predictably Irrational: the Hidden Forces That Shape Our Decisions. Harper Perennial, 2010.

Chapter 5 & 6:

Alpaydin, Ethem. Introduction to machine learning. MIT press, 2020.

Probably approximately correct: nature's algorithms for learning and prospering in a complex world. Basic Books (AZ), 2013.

Shalev-Shwartz, Shai, and Shai Ben-David. Understanding machine learning: From theory to algorithms. Cambridge university press, 2014.

Vapnik, Vladimir N. "An overview of statistical learning theory." IEEE transactions on neural networks 10.5 (1999): 988-999.

Murphy, Kevin P. Machine learning: a probabilistic perspective. MIT press, 2012.

Goodfellow, Ian, Yoshua Bengio, and Aaron Courville. Deep learning. MIT press, 2016. Valiant, Leslie.

Chapter 7:

Pauen, Sabina, and Stefanie Hoehl. "Preparedness to learn about the world: Evidence from infant research." Epistemological dimensions of evolutionary psychology. Springer, New York, NY, 2015. 159-173.

Seligman, Martin EP. "Phobias and preparedness." Behavior therapy 2.3 (1971): 307-320.

Kotseruba, Iuliia, and John K. Tsotsos. "40 years of cognitive architectures: core cognitive abilities and practical applications." Artificial Intelligence Review 53.1 (2020): 17-94.

R. A. Brooks, "Intelligence Without Representation", Artificial Intelligence 47 (1991): 139-159.

R. A. Brooks, "A Robust Layer Control System for a Mobile Robot", IEEE Journal of Robotics and Automation RA-2 (1986): 14-23.

Russell, Stuart. Human compatible: Artificial intelligence and the problem of control. Penguin, 2019.

Ghahramani, Z. "Probabilistic machine learning and artificial intelligence." Nature 521 (2015):452–459.

Ellis, Kevin, Armando Solar-Lezama, and Josh Tenenbaum. "Unsupervised learning by program synthesis." (2015).

Lake, Brenden M., Ruslan Salakhutdinov, and Joshua B. Tenenbaum. "Human-level concept learning through probabilistic program induction." Science 350.6266 (2015): 1332-1338.

Lerer, Adam, Sam Gross, and Rob Fergus. "Learning physical intuition of block towers by example." International conference on machine learning. PMLR, 2016.

Ullman, Tomer D., et al. "Learning physical parameters from dynamic scenes." Cognitive psychology 104 (2018): 57-82.

Botvinick, Matthew, et al. "Building machines that learn and think for themselves." *Behavioral and Brain Sciences* 40 (2017).

Ramsauer, Hubert, et al. "Hopfield networks is all you need." arXiv preprint arXiv:2008.02217 (2020).

Chapter 8:

Minsky, Marvin. Society of mind. Simon and Schuster, 1988.

Baum, Eric B. "Toward a Model of Mind as a Laissez-Faire Economy of Idiots." ICML (1996).

Simon, Herbert A. "The Architecture of Complexity." *Proceedings of the American Philosophical Society*, vol. 106, no. 6, American Philosophical Society (1962).

Chang, Michael, et al. "Decentralized reinforcement learning: Global decision-making via local economic transactions." International Conference on Machine Learning. PMLR, 2020.

Kwee, Ivo, Marcus Hutter, and Jürgen Schmidhuber. "Market-based reinforcement learning in partially observable worlds." International Conference on Artificial Neural Networks. Springer, Berlin, Heidelberg, 2001.

Dai, Xiaowu, and Michael Jordan. "Learning in Multi-Stage Decentralized Matching Markets." *Advances in Neural Information Processing Systems* 34 (2021).

Malik, Momin M. "A hierarchy of limitations in machine learning." arXiv preprint arXiv:2002.05193 (2020).

D. Acemoglu, M. I. Jordan, and E. Glen Weyl. "The Turing Test is bad for business: Technology should focus on the complementarity game, not the imitation game." WIRED Magazine, 2021.

Chapter 9:

Koch, Christof. "Biophysics of Computation: Information Processing in Single Neurons". Oxford University Press, 1998.

Koch, Christof, and Idan Segev. "The role of single neurons in information processing." Nature neuroscience 3.11 (2000): 1171-1177.

Friston, Karl. "The free-energy principle: a unified brain theory?" Nature reviews neuroscience 11.2 (2010): 127-138.

Scellier, Benjamin, and Yoshua Bengio. "Equilibrium propagation: Bridging the gap between energy-based models and backpropagation." *Frontiers in computational neuroscience* 11 (2017): 24.

Bargh, J. A., & Chartrand, T. L. (1999). The unbearable automaticity of being. American Psychologist, 54(7), 462–479. Evans, Richard, et al. "Making sense of sensory input." Artificial Intelligence 293 (2021): 103438.

Zobisch, Paula J. The theory of multiple intelligences and critical thinking. Diss. Capella University, 2005.

Hawkins, Jeff, et al. "A framework for intelligence and cortical function based on grid cells in the neocortex." Frontiers in neural circuits 12 (2019): 121.

Alexander, Stephon, et al. "The Autodidactic Universe." arXiv preprint arXiv:2104.03902 (2021).

Appendix A: Generalization of the Missing Manual Problem

As we mentioned in Chapter 3, we need a way to talk about science and philosophy without specifying which science or which philosophy in particular! This seems like a hopelessly general and abstract task. But we don't have a choice as we do need a general framework to discuss foundations for a science that doesn't yet exist! It's not even clear what we mean by foundations. Therefore, we may not be able to aim for precision or tangibility yet, but we could go for conceptual enormity.

In order to prevent our analysis throughout the book from getting uselessly complicated, let us develop a metaphor to fall back on for simplicity. It must be a metaphor that is generally applicable. Simply because not only many different disciplines contributed to the birth of AI but also many may have to recombine later to give rise to a proper science of intelligence. This unknown necessitates adopting a generally applicable metaphor. The language we adopt should also give us a way to represent both philosophy and science as we need both to talk about proper foundations, as we have argued earlier.

A great candidate is the previous chapter's metaphor of a missing manual regarding our minds and thinking. We can simply generalize that notion and use it throughout the book.

A deep problem, by definition, always has many consequences manifesting themselves as different sets of problems. Until clear-cut consequences rise up to the surface of our attention, we don't typically get to identify them. We see the fruits, not the first seed. The only field which aims at being aware of the problems irrespective of the consequences or whether they are fully visible or measurable, is philosophy. So, let's remind ourselves what philosophy is.

Philosophy studies the nature of knowledge, whether it's inside us, outside, or inside another mind, as well as how they all relate, what can be known, what can't, and so forth. For instance, while science may say what's possible with a high degree of certainty, philosophy concerns itself with what science can possibly know and also what it means to actually know it. Most importantly, many consider the job of modern philosophy to be relating different kinds of knowledge to one another in a coherent manner. In particular, how scientific knowledge maps to our knowledge of our everyday experiences. This is definitely not a trivial task, to say the least.

What we discussed in the last chapter has lots of overlaps with what the philosophy of mind tries to address too, which includes consciousness, cognition, etc. However, if there was no missing manual problem there would hardly be a need for such areas of philosophy to exist. So there is some causal relationship at play which we can exploit to redefine the missing manual problem more concisely.

Let us first extend the notion of the missing manual problem beyond that of the last chapter, which was specifically tied with mental phenomena. We could have equally well-considered any other phenomena. Think of an organizational or social phenomenon for example. To every phenomena and domain (which may permit its own sets of philosophies) we could associate some other imaginary (and therefore missing) manual that contains ultimate words or descriptions of that phenomena, like we associated one with our own minds in Chapter 2. Suppose we've done that. Now, we've got a multitude of missing manual problems as we are concerned not just with ourselves but nature at large within which we're deeply embedded. In doing so, we have to stretch the notion of a manual to be not just a property of machines but also any phenomena. So from here on, by manual, we mean this very generalized notion.

Recognizing that this generalization is perhaps too much of a stretch load on the word "manual" (which implies some instructions to operate some machine that exists a priori), how about we invent a single new word to refer to this generalized notion. Let's call it a **Manuon**, like a "manual of nature". Manuon would be the manual that nature "uses" to be the way it is, or gives rise to any phenomena that she does.[169] It includes facts and patterns in any given target domain. Wherever we could say nature operates in different domains, we could also say there exist different manuons, again metaphorically.

---

[169] Though we have lots to talk about the word nature or natural. Here we are not suggesting any philosophy of nature whatsoever, this is simply a choice of language, and a very crude one.

The literal translation of philosophy is "to love wisdom" and the goal of philosophy is generally considered "to be wise"! On the other hand, by definition, the missing manual is supposed to contain and state the "wise way of conduct". Therefore, a part of any missing manual can be considered to be an imaginary answer sheet to some particular domain of philosophy, the domain associated with the definition of that manual. So each missing manual covers some domain of philosophy. But it also contains specific instructions on how to do something and how it works beyond what concerns philosophy. This is what the science of that domain aims to obtain. For instance, using the metaphor of machines, the manual tells you what "buttons" are there (by analogy to the actual experiments you could possibly run), what you should get (results of those experiments), the limits, etc.

The fact that such a hypothetical manual for some domain X is missing, is what causes there to be a philosophy and science for that domain. This brings science and philosophy under one roof making it more obvious that we need both. In other words:

The missing manual of domain X can be used as a container for both the science and philosophy of X.

On the flip side, the practice of science and philosophy can be imagined to be the study of different phenomena in order to put together some *artificial* **manual**. Physics for instance attempts to discover/describe how nature works. That could be thought of as finding a copy of the laws of nature, whatever nature follows, whatever one may consider written in its "manual". If you put all those "laws" next to each other and label the collection, the manual for nature, then theoretical physics can be thought of as building an artificial version of that manual.[170]

Obviously the same goes for other sciences. Therefore,

Science and philosophy of domain Y, output an Artificial Manual of Y.

---

[170] An ideal manual covers all sorts of conditions and regimes, so it doesn't mean that the MM of physics just contains absolute facts. So things like Newton's laws for instance do have a place in it. Every successful theory has its own utility and domain of applicability. In other words, the MM may be considered and allowed to be full of "effective" theories.

So effectively we have made some "change of variables". Instead of science and philosophy, we talk about missing manuals and artificial manuals; cause and effect pairs. The table below captures this change of variables and the usage of the metaphor in the book:

Science & Philosophy (S&Ph)	*(imaginary) Manual/Manuon*
Cause for a domain of S&Ph to exist.	A missing manual.
Result/product of a domain of S&Ph.	An artificial manual.
An independent domain of S&Ph.	A single self-contained artificial manual.
Philosophy of a domain of S&Ph.	Discussion of the artificiality of an artificial manual and how it should/will evolve.
Foundations for a domain of S&Ph.	The justifications for the outline and definitions of the manual.
Proper foundations for a domain of S&Ph.	If there exists only one artificial manual which you can build upon and extend its outline and definitions without changing the existing outline and definitions.

TABLE 1

This crude metaphor is nothing but a "change of variables" in talking about science and philosophy. Change of variables cannot make subjectiveness disappear. It remains subjective as to what we consider as foundations, and what is exactly considered a domain with proper foundations. Without getting into the specifics of any field, its history, and its goals, it's rather impossible to become more objective about the topic of foundations and measures of effectiveness. So unfortunately, that will remain vague until we take the specifics of any given discipline into account. That's what we do with the science of cognition in chapter 3 and with AI in later chapters.

Having said that, we have achieved what we wanted for now. We wanted a simplifying metaphor that 1) is generally applicable, 2) doesn't separate science and philosophy. Both are obviously check-marked. The accessible language we are adopting is obviously and by-construct generally applicable. It also allows us to ignore any separation of science and philosophy. What further we can appreciate is that a vocabulary based on artificial manuals makes it explicit that we are never done in science and philosophy and it's always a work in progress. It enables us to talk more naturally about the dynamics of how science and philosophy evolve rather than what they may say at any given moment in time and that we argue in the next chapter to be a blind spot in current philosophies of science.

In summary, the position we're taking in this book is set by our philosophy regarding the evolution of any discipline. That is to metaphorically imagine there is a missing manual that nature holds and it contains the ultimate philosophies, truth, useful facts, and recipes regarding the subject under consideration. And that our job can be interpreted as putting together an artificial version of the "natural" manual that is missing, and gradually get the artificial manual closer to the "natural" one. Think of the natural one as what "God would put together" and the artificial one being our collective knowledge at any given time. We also- swapped the word manual for the more suitable MANUON (manual of nature or natural manual) in order to drop some of the inappropriateness that the word manual carries in its literal sense.

Appendix B: Sets, Functions, and Maps

Sets

A set is a collection of things. The important thing to note is that the set itself is one separate thing. A set has no structure, it just says whether an object is part of a set or not, it's only a statement about existence. For that reason, sets are the most primitive objects in all of mathematics.

Functions

Consider the people in our world, everybody has a name, and then suppose we all could enter a different world in which all our names get inverted. If I am called "Ali" here, there I'd be "ilA". This is called mapping, names in this world get mapped to names in the new world. Everything in the world gets matched with a counterpart in a different world. They are not the same thing, there just exists a MAPPING between them. A function is a mapping where every member of the old world gets mapped to only one counterpart in the new world. If no one in the old world is left behind, meaning everyone gets matched with some counterpart in the new world then that mapping is called an *entire function*, otherwise, it's a *partial* function because some of the members of the old world don't get to have a match.

Functions map every member of the input set (the domain of the function) to one and only one member of the output set. Given that sets are so primitive, and functions are what take us from one set to another, you may wonder what is it that sets and functions cannot represent for us? In principle, nothing. Therefore, if we chose to see the world through sets and functions, we can certainly see them everywhere. So, if you have a good function approximation method (as we do with deep learning), it is like having a hammer in a sea of nails! We are highly tempted to build everything with nails except that if we build too big a structure with nails it would certainly fall apart!

Just like sets are the most primitive objects, functions are also the most primitive maps in mathematics. The fact that everything can be seen as sets and functions doesn't mean such representations are always convenient, suitable, or even helpful at all. For that reason, in mathematics, we have plenty of other objects and maps, beyond sets and functions.

As we said, sets are structureless, although the structure could be hidden inside an element of a set if the set is made of rich objects. Examples of such objects are manifolds (think of them as locally smooth spaces) or categories (think of them as structures of mathematical objects and relations). They get transformed by maps that care about the structure, such as *continuous maps* that are used in topology, or *functors* that are used to map categories to one another.

Structure-preservation is not the only thing that escapes functions. There are many other subtleties that require more sophisticated maps than functions. Functions of functions, sometimes called higher-order functions are one example. Sets of functions that are related by a property or constraint, known as distributions, are another. Distributions allow us to deal with many subtleties involving ill-defined functions. Consider a bizarre function that is zero everywhere in its domain except for one point for which it assigns infinity, the so-called Dirac delta-function which is, in fact, not a function but the limiting case of a distribution of functions. Distributions are generalized functions which require assistance from some other (so-called test) functions that are well-behaved enough compared to a target function (called the core of the generalized function) to define themselves as well-behaved maps. Statistical distributions are only special cases of generalized functions (mathematical distributions) and most often are just regular functions satisfying some properties (probability axioms).

Maps

Mapping is not a single formal concept in mathematics unlike the concept of a function. The reason is that basically anything and everything in mathematics can be understood as an act of mapping, going from one (or set of) mathematical object(s) to another (set of) object(s). This is also why we often use the word transformation in place of "mapping". If it sounds overly broad it is because it is. Everything in mathematics like equations, inequalities, logical or set-theoretic statements, etc. are all mappings. Even stating what *a set* is, introducing it, is itself a map. And if you believe that everything we talk about in any language (natural or artificial) we have ever constructed, has a mathematical reflection, then any thought you may ever have is also an act of mapping. You go from something to something else. What's in between? Just a map. All of mathematics is a series of maps that can be constructed from some very primitive sets. From that, we can get to the most sophisticated maps of maps. Set aside disputes in the foundations of mathematics about which sets to start from and which axioms, those don't change the fact that all that is happening is mapping.

OK. everything is a map, so you say. What is NOT a map? Good question. Note that we are not saying that everything *is* a set or a map or boils down to a set or a map. We are not making any ontological statements here. We are just saying that sets and maps are the most fundamental concepts in mathematics, as a representation of things. We are not yet talking about the nature of reality, and what is really out there. It may be that the fundamental bias of our brain (see Chapter 3) is at work for our perception of reality to see maps as the ultimate expressions of nature. This is not any different than to say, "mathematics is the language of nature". At any rate, we had to establish these facts for the following reasons:

1) We do have to discuss the nature of reality at some point. Coming later!

2) We are reminding ourselves that everything is a map, and functions as the most primitive of all maps are also everywhere. Acknowledging that, of course, we can continue to laser focus on approximating as many functions as we can, each with a separate deep learning model. But even after we discover, in our deep learning research, how to approximate many kinds of functions that we don't yet know how to approximate, there will always be more relevant (but not new kinds of) functions out there to approximate in practice. Our never-stopping hunger for doing/predicting more things (and better) would then mean we are bound to hit a ceiling with technologies that are nothing, but a collection of approximated functions glued together with some rigid application logic.

3) We are reminding ourselves that mathematics is not stuck at the level of sets and functions, just because they are in principle general enough. Convenience and efficiency of representation have always mattered. By analogy, we can suspect that AI cannot and should not stop at function approximation either. We may want to work with more sophisticated techniques such that we can manage the relationships between many functions, higher-level structures, and properties that may emerge from the interaction of such functions and structures. Perhaps we should expect the future of AI to get more *mathy*!

Acknowledgements

I'd like to thank Manuel Aparicio, Sidharth Thakur, Simone Sartoretto, Daniela de Albuquerque, Evandro Barbosa, and James Fleming for their review of the manuscript and feedback.